THEORIES OF CHEMICAL
REACTION RATES

THEORIES OF CHEMICAL REACTION RATES

KEITH J. LAIDLER

Professor of Chemistry
The University of Ottawa

McGRAW-HILL BOOK COMPANY

New York   St. Louis   San Francisco   London
Sydney   Toronto   Mexico   Panama

THEORIES OF CHEMICAL REACTION RATES

Library of Congress Catalog Card Number 78-85165

35834

234567890 MAMM 765432

935953

Preface

The problem of interpreting and predicting the rates of chemical reactions has proved to be enormously difficult. The very simple hard-sphere collision theory of reactions was widely used at one time, but was soon found to be unsatisfactory. In the 1930s the development of activated-complex ("absolute reaction rate") theory represented a very important achievement. Despite some obvious limitations, this theory gave considerable insight into the way in which chemical reactions occur and has provided a basis for calculating the rates of various kinds of chemical and physical processes.

Until a few years ago there was little reason to explore alternative theories of reaction rates since few experimental results were available which would permit one to discriminate between alternative theories. However, with the recent development of techniques for carrying out chemiluminescence and molecular-beam studies a body of experimental data has become available which allows a more detailed probing into the mechanisms of elementary processes. This has led to the formulation of alternative theories, of which those classed as "dynamical" are the most useful; these treat molecular collisions in a much more realistic way than had previously been possible.

This book gives a general account of the various theories that have been developed to explain reaction rates. A considerable part of the book is concerned with activated-complex theory because of the wide applicability of this theory; the dynamical and other, more detailed, theories are also discussed. In all cases an effort has been made to indicate the applicability of the theories to the experimental data.

This book is based on a course of lectures on rate theory that I gave at four universities in England during my tenure of a Commonwealth Visiting Professorship in the academic year 1966–1967. I am grateful to the Association of Commonwealth Universities for arranging this professorship and to the staff of the following universities, where the lectures were given, for kind hospitality and helpful discussions: The University of Sussex; Queen Mary College, London; the University of Kent at Canterbury; and the University of Hull.

I am also grateful to a number of colleagues who have made valuable comments on the manuscript, particularly to Dr. Margaret H. Back (University of Ottawa), Professor Dudley Herschbach (Harvard University), Dr. K. A. Holbrook (University of Hull), Professor David M. Bishop (University of Ottawa), Professor John C. Polanyi (University of Toronto), and Dr. P. J. Robinson (University of Manchester).

KEITH J. LAIDLER

Contents

THEORIES OF CHEMICAL
REACTION RATES

1
Introduction

The understanding of chemical reaction rates is one of the most important problems in chemistry. It is necessary to gain a knowledge, by means of theories, of the main factors that influence reaction rates. It is also important to be able to make calculations, from first principles, of the rates of at least the simpler types of reactions. Neither of these two objectives has yet been satisfactorily attained. Our understanding of the factors influencing rates is still based heavily on empiricism, and the calculation from first principles of the rate of even the simplest chemical reaction has still not been accomplished. This book is concerned with the main theories which have been developed in the attempt to deal with these two problems.

There are two main theoretical aspects of reaction rates, which arise as a result of Arrhenius' formulation of the rate constant as

$$k = Ae^{-E_{exp}/RT} \tag{1}$$

The preexponential or frequency factor A and the activation energy E_{exp} are practically independent of temperature. The calculations that have been made of activation energies are considered in some detail in the next chapter. We shall see that the purely quantum-mechanical methods are slowly becoming sufficiently refined that a reliable activation energy can be calculated for the simplest of reactions, but that the prospect for more complex reactions is still not at all encouraging. Much greater success is achieved if some degree of empiricism is permitted, as is also considered in the next chapter.

The frequency factor A has also presented considerable theoretical difficulty. The first attack on the problem was made by Trautz[1] and W. C. McC. Lewis,[2] who identified A with a collision number which they calculated by using a simple theory of collisions in which the molecules were treated as structureless spheres. This theory is certainly along the right lines, but the hard-sphere concept is too crude and leads to serious numerical errors for many kinds of reactions. An alternative treatment, worked out independently by Eyring[3] and by Evans and Polanyi,[4] is also a collision theory, but it deals with collisions in a more satisfactory way by taking into account the structures of the reacting molecules and the manner in which the reacting molecules come together on collision. These formulations focus attention on a particular species, known as the *activated complex*, which is in the act of passing over the top of the potential-energy barrier. It is assumed in these theories that the concentrations of these activated complexes can be calculated on the basis of equilibrium theory, an assumption that introduces a great simplification into the formulation of the rate equation. Under certain conditions, which apply to most of the reactions that have been studied and indeed to most chemical reactions of the conventional kind, this equilibrium assumption can be justified. Chapter 3 is concerned with these formulations of the rate equation and discusses the equilibrium assumption in some detail.

Figure 1 presents a classification of the various theories of reaction rates. The first subdivision is into whether or not the theories explicitly consider potential-energy surfaces, which are the "maps" that relate potential energy to the geometry of the reactant molecules and the reaction intermediates. A theory capable of providing rates for particular reactions must in some way take account of these potential-energy variations, but a number of more formal treatments of rates, although not providing numerical values for rates, have been valuable in leading to useful conclusions; examples are the stochastic (random-walk) theories and rate theories based on nonequilibrium statistical mechanics.

[1] M. Trautz, *Z. Anorg. Allgem. Chem.*, **96**:1 (1916).

[2] W. C. McC. Lewis, *J. Chem. Soc.*, **113**:471 (1918).

[3] H. Eyring, *J. Chem. Phys.*, **3**:107 (1935).

[4] M. G. Evans and M. Polanyi, *Trans. Faraday Soc.*, **31**:875 (1935).

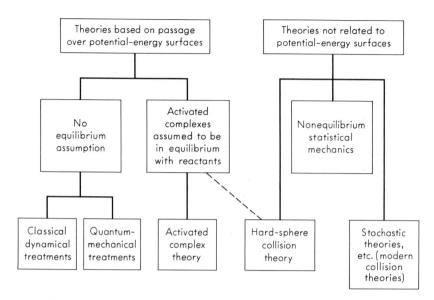

Fig. 1 Theories of chemical reaction rates.

Theories may also be classified according to whether or not they make the assumption of equilibrium between reactants and activated complexes. The main equilibrium theory that is related to potential-energy surfaces is the activated-complex theory of Eyring and of Evans and Polanyi. A number of recent theories have avoided the equilibrium assumption and have treated, by either classical or quantum mechanics or by a combination of the two, the passage of systems over potential-energy surfaces. These theories are necessarily much more complicated than the equilibrium theories, and it is difficult to obtain from them numerical values which can be compared with experimental results.

There are several reasons why it is important to have nonequilibrium theories. In the first place, the validity of the equilibrium assumption can be questioned, and it is desirable to have an alternative formulation. In particular, when a reaction is very fast there is a depletion of the more active reactant molecules, and then there is bound to be some error in assuming the activated complexes to be in equilibrium with the reactants. Apart from these considerations, there are many situations where one is interested in understanding the detailed molecular dynamics of a chemical process. For example, in connection with molecular-beam experiments and with studies of the quantum states of product molecules, one needs a more precise treatment of the collision processes than can be provided by the theories in which the equilibrium assumption is made. Nonequilibrium theories are considered in Chaps. 7 and 8.

THE ARRHENIUS LAW

All theories of chemical kinetics bear a close relationship to the Arrhenius law [Eq. (1)], which was originally an empirical and macroscopic law relating the rate constant to the temperature.   The great virtue of this law is its simplicity; it involves only two parameters, the *experimental activation energy* E_{exp} and the *preexponential* or *frequency*[1] *factor* A.   It is remarkably satisfactory, few experimental results being of sufficient accuracy that significant deviations from the law can be detected.   However, the precise meaning of the parameters E_{exp} and A is by no means simple and straightforward.

Arrhenius[2] and van't Hoff[3] originally interpreted the experimental activation energy as simply the height of the energy barrier which has to be overcome in order for reaction to occur.   This concept is basically correct, but is oversimplified since we now know that the probability of reaction is a rather complex function of the various kinds of energy in the reacting molecules and also of the relative configuration of the reactant molecules when the collision takes place.   The various attempts to take this into account, some of which are dealt with in later chapters, lead to equations of the type

$$k = B(T)e^{-E_0/RT} \tag{2}$$

where E_0 is the threshold energy, i.e., the lowest relative translational energy at which reaction can occur.   The preexponential factor $B(T)$ is now temperature-dependent, its form depending on the specific assumption made in the theory about which part of the total energy of the system contributes toward reaction.   Since this depends upon the nature of the reaction, no single version of Eq. (2) is likely to be widely applicable. Moreover, since $B(T)$ is usually much less temperature-dependent than the exponential term, it has proved very difficult to obtain reliable experimental evidence of the temperature dependence of $B(T)$.   Indeed the simple Arrhenius equation (1), with the preexponential factor assumed constant, has usually been found to fit the results at least as well as any of the modified equations.

The reaction cross section[4]   Particularly for the interpretation of the more recent work on the kinetics of bimolecular reactions it has been found

---

[1] The term *frequency factor*, although very commonly employed, is not a particularly happy one since its dimensions are not necessarily those of a frequency.   Its units are the same as for k; for example, seconds^{-1} for a first-order reaction, cubic centimeters mole^{-1} sec^{-1} for a second-order reaction.

[2] S. Arrhenius, *Z. Physik. Chem. (Leipzig)*, **4**:226 (1889).

[3] J. H. van't Hoff, "Etudes de dynamique chimique," F. Muller and Company, Amsterdam, 1884.

[4] Cf. E. F. Greene and A. Kuppermann, *J. Chem. Ed.*, **45**:361 (1968).

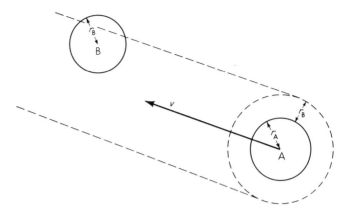

Fig. 2 The frequency of collisions. A is moving with velocity v relative to B and can be regarded as surrounded by a sphere of radius $r_A + r_B$. In 1 sec this sphere has swept out a volume of $\pi(r_A + r_B)^2 v$ and has encountered $\pi(r_A + r_B)^2 v n_B$ molecules of B; the frequency of collisions of n_A molecules that are present in 1 cc is therefore $\pi(r_A + r_B)^2 v n_A n_B$ molecules cc^{-1} sec^{-1}.

convenient to make use of the concept of the *reaction cross section* σ. On the basis of elementary kinetic theory (Fig. 2) the number of collisions per second in 1 cc between n_A molecules of A and n_B of B, approaching one another with relative velocity v, is

$$\text{No. of collisions} = \pi(r_A + r_B)^2 v n_A n_B \qquad \text{molecules cc}^{-1}\text{ sec}^{-1} \qquad (3)$$

where $r_A + r_B$ is the sum of the radii. If every collision led to reaction, this would be the rate; in general we can express the rate in terms of the *reaction cross section* σ, using the analogous equation

$$\text{Rate} = \sigma(v) v n_A n_B \qquad (4)$$

The second-order rate constant k_v for reaction between A and B moving at relative velocity v is defined by

$$\text{Rate} = k_v n_A n_B \qquad (5)$$

The rate constant k_v is thus

$$k_v = \sigma(v) v \qquad (6)$$

If the reacting species possess a distribution of velocities $f(T,v)$, the rate constant (6) has to be weighted by this function to give the average rate constant

$$k(T) = \int_0^\infty f(T,v)\sigma(v) v \, dv \qquad (7)$$

For a system at thermal equilibrium the function $f(T,v)$ corresponds to the Maxwell-Boltzmann distribution, and it is then found[1] that

$$k(T) = \left[\left(\frac{2}{kT}\right)^{3/2}\left(\frac{1}{\mu\pi}\right)^{1/2}\int_0^\infty E\sigma(E)e^{(E_0-E)/RT}\,dE\right]e^{-E_0/RT} \tag{8}$$

where μ is the reduced mass.

The activation energy   The experimental activation energy $E_{\exp}$ is by definition calculated from the slope of the plot of log k against $1/T$ and is given by

$$E_{\exp} = -R\frac{d\ln k(T)}{d(1/T)} = RT^2\frac{d\ln k(T)}{dT} \tag{9}$$

If this equation is combined with (8), we obtain

$$E_{\exp} = \frac{\int_0^\infty \sigma(E)E^2 e^{-E/RT}\,dE}{\int_0^\infty \sigma(E)E e^{-E/RT}\,dE} - \tfrac{3}{2}RT \tag{10}$$

This relationship was first derived by Tolman.[2]   The first term on the right-hand side of Eq. (10) is the average energy $\bar{E}^*$ of those collisions in which reaction takes place, and $\tfrac{3}{2}RT$ is the average energy (per mole) for all collisions; the latter may be written as $\bar{E}$.   The activation energy is thus simply

$$E_{\exp} = \bar{E}^* - \bar{E} \tag{11}$$

This relationship is illustrated by the curves in Fig. 3, which were calculated by Menzinger and Wolfgang.[3]   The diagram relates to a threshold energy E_0 of 8.0 kcal/mole and to a temperature of 300°K. The left-hand curve is the ordinary Maxwell-Boltzmann distribution curve for 300°K; the average energy $\bar{E}$ is indicated.   For the purpose of these calculations it was assumed that the cross section $\sigma(E)$ is proportional to $(E - E_0)^2$.   The curve designated *reaction function* was calculated by taking, at a range of energies, the product

$$f(T,v)\sigma(v)v$$

as in Eq. (7); this quantity is the energy distribution of collisions resulting in reaction.   The average value of this function, $\bar{E}^*$, is indicated on the diagram; like $\bar{E}$ it does not correspond to the maximum in the curve.

[1] M. Eliason and J. O. Hirschfelder, *J. Chem. Phys.*, **30**:1426 (1959); M. Karplus, R. N. Porter, and R. D. Sharma, *J. Chem. Phys.*, **43**:3259 (1965).

[2] R. C. Tolman, *J. Am. Chem. Soc.*, **42**:2506 (1920).

[3] M. Menzinger and R. Wolfgang, *Angew. Chem.*, April, 1969.

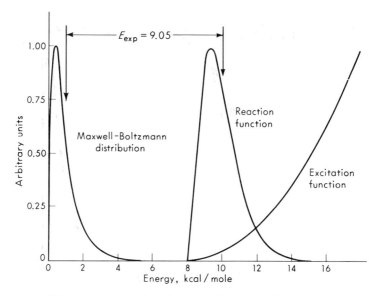

Fig. 3 Distribution functions for a reaction at 300°K with a threshold energy E_0 of 8.0 kcal.

For the particular excitation function chosen the experimental activation energy E_{exp}, equal to 9.05 kcal at 300°K, is significantly higher than the threshold energy E_0. As the temperature increases, the change in $\bar{E}^*$ will not in general equal the change in E, so that E_{exp} is temperature-dependent; thus at 900°K the calculations for this case lead to an E_{exp} value of 11.8 kcal/mole. Other forms for the excitation function lead to different variations of E_{exp} with temperature; the matter is considered in some detail by Menzinger and Wolfgang. Usually it is found that E_{exp} does not vary strongly with temperature; in other words, the Arrhenius law, although not perfect, is a very good approximation.

Similar considerations lead to conclusions about the temperature dependence of the preexponential factor.

The above arguments assume an equilibrium distribution of reactants. The effect of chemical reaction is to remove the more energetic molecules, so that the Maxwell-Boltzmann distribution will not be maintained. This effect is of relatively greater importance for rapid reactions and has been dealt with theoretically in a number of ways to be considered in later chapters.

2
Potential-energy Surfaces

It was realized by Arrhenius[1] that the relationship between the rate constant of a reaction and the temperature leads to the conclusion that there is an energy barrier to reaction. In most cases reaction does not invariably occur when two reactant molecules collide with one another, even if the relative orientations are suitable; the molecules must possess between them a certain critical amount of energy.

It was first suggested by Marcelin[2] that the activation energy is conveniently treated in terms of the concept of a *potential-energy surface* or hypersurface, which results when potential energy is plotted against suitable bond distances and angles. If the reaction is between two atoms, only one distance, that between the nuclei, is involved, and one can plot potential energy against this distance; the result, a two-dimensional dia-

[1] S. Arrhenius, *Z. Physik. Chem. (Leipzig)*, **4**:226 (1889); see M. H. Back and K. J. Laidler, "Selected Readings in Chemical Kinetics," p. 31, Pergamon Press, Oxford, 1967.

[2] A. Marcelin, *Ann. Phys.*, **3**:158 (1915).

gram, is the potential-energy *curve* for the diatomic molecule. If three atoms are involved, as in a reaction of the type

A + B—C → A—B + C

the system A····B····C must be described in terms of three parameters; these might be the A—B, B—C, and A—C distances or the A—B and B—C distances and the angle A—B̂—C.   In order to plot energy against these three distances, a four-dimensional diagram would be necessary. Since such a diagram cannot be constructed or visualized, it is necessary to use a series of diagrams in each of which one parameter has been fixed at a particular value.   For example, the A—B̂—C angle might be fixed at 180° and a three-dimensional model constructed.   Other three-dimensional surfaces could be constructed for other angles, and all of these would be sections through the four-dimensional surface.

When this procedure of fixing the A—B̂—C angle is adopted, the resulting potential-energy surface has the form shown in Figs. 4, 5, and 6. On the left-hand face of Fig. 4 the B—C distance may be considered sufficiently great that one is dealing simply with the diatomic molecule A—B; the curve shown on that face is thus the potential-energy curve for the molecule A—B, the point R corresponding to the dissocation of the molecule and the point Q to the classical ground state.   Similarly,

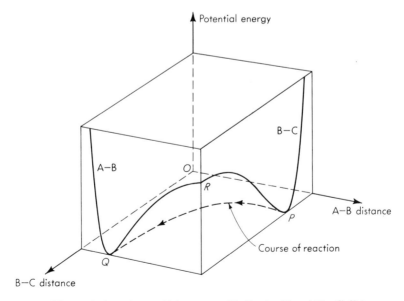

Fig. 4   The variation of potential energy with the A—B and B—C distances, for the A····B····C system in which the A—B—C angle has been fixed.

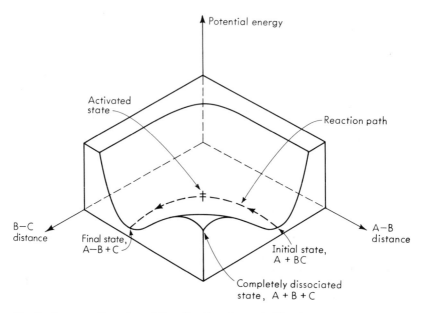

Fig. 5 A perspective view of the situation shown in Fig. 4.

on the right-hand face of the diagram there is a curve for the diatomic molecule B—C, the distance A—B now being sufficiently great that A has no effect on the energy of the system.   The classical ground state of B—C is represented by the point P.

The course of the reaction is represented by a movement on the potential-energy surface from P to Q, and the system will tend to travel along paths where the energy is not too high.   The calculations have shown that the general form of a potential-energy surface of this kind is that, running in from the points P and Q on the diagram, there are two valleys which meet in the interior at a *col*, or *saddle point*.   This is seen in the perspective drawing shown as Fig. 5 and in the contour diagram shown in Fig. 6.   In order for the system to pass from P to Q it will tend to travel along the bottom of the first valley, over the col, and down into the second valley.   The *minimum-energy reaction path*, corresponding to the bottoms of the valleys, is shown in Fig. 6 as a dashed line.   The paths, or *trajectories*, followed by individual reactions will be not too far from this minimum path, since paths that are too far away from it will involve too great an expenditure of energy.

A section through this minimum path, known as a *reaction profile* or *potential-energy profile*, is shown in Fig. 7.   Particular significance attaches to the point at the top of this profile; this point corresponds to

the saddle point in Figs. 4 to 6.   This point is not only a position of
maximum energy with respect to the reaction path, it is also a position
of minimum energy with respect to motions at right angles to the reaction
path.   Systems corresponding to a small area around this point are known
as *activated complexes* and are denoted by the symbol ‡; as will be seen
later (page 49), there is arbitrariness associated with the definition of an
activated complex.

Potential-energy surfaces for reactions involving more than three
atoms are hypersurfaces in multidimensional space.   In practice they
can only be constructed by holding constant all but two variables.
Alternatively, one may, as an approximation, construct surfaces for
more complicated systems by regarding certain groups of atoms as single
particles.   For example, potential-energy surfaces have been constructed
for the reaction

$$Na + CH_3I \rightarrow Na + CH_3I$$

by treating the CH_3 radical as a single particle (see Fig. 71, page 192).

The construction of potential-energy surfaces is obviously a matter

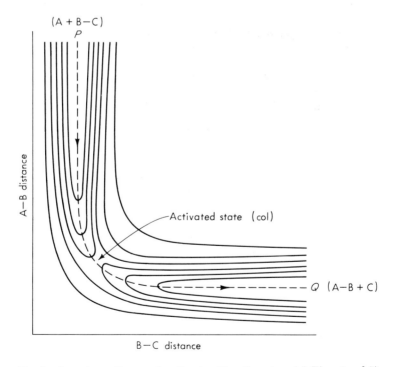

Fig. 6   A contour diagram for the A····B····C system (cf. Figs. 4 and 5).
The dotted line shows the minimum-energy path.

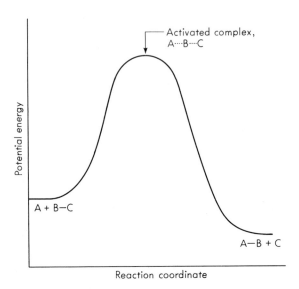

Fig. 7 A section through the minimum path shown in Fig. 6; this section is known as a potential-energy profile.

of very considerable importance. Essentially the problem is a quantum-mechanical one; one needs to calculate, for a whole series of A$\cdots$B$\cdots$C configurations, the potential energy, and in this way build up the complete surface. This matter can readily be dealt with in principle, but in practice it is exceedingly difficult, by purely quantum-mechanical procedures, to calculate energies that are sufficiently accurate to be acceptable for kinetic purposes. This is true even for the simplest system, H$\cdots$H$\cdots$H, which contains only three electrons; for systems containing more electrons, the errors that arise when the calculations are made are much more serious.

Because of the difficulty of making purely quantum-mechanical calculations, a good deal of effort has gone into the construction of potential-energy surfaces by methods into which a good deal of empiricism has been introduced. Such methods are referred to as *semiempirical* and *empirical* treatments. There is no sharp line separating semiempirical from empirical treatments; the former rely more on purely theoretical methods, while the latter are only slightly based on theory and lean heavily on experimental generalizations. By and large the purely empirical treatments are procedures for interrelating the experimental results, while the semiempirical treatments have a theoretical basis but involve adjustments on the basis of the results they are designed to explain. The remainder of this chapter is concerned with these purely quantum-mechanical, semiempirical, and empirical procedures for constructing potential-energy surfaces.

PURELY QUANTUM-MECHANICAL CALCULATIONS

The quantum-mechanical calculations are based on the Born-Oppenheimer approximation, according to which electronic energy can be considered at a particular internuclear separation. This means that it is possible to consider the nuclei fixed at a certain separation and then to calculate the potential energy. This procedure can then be repeated for a variety of nuclear configurations. Provided that this approximation holds and provided also that the electrons remain in one state (the *adiabatic* assumption, see page 57), a single potential-energy hypersurface can be constructed to give the potential energy E for all internuclear separations of an assembly of atoms.

Many of the calculations that have been made have related to the reaction

$$H^\alpha + H^\beta\!\!-\!\!H^\gamma \to H^\alpha\!\!-\!\!H^\beta + H^\gamma$$

The rate of this reaction can be studied either by using D in place of H (in which case there is a complication due to a kinetic-isotope effect) or by measuring the rate of the para-ortho hydrogen conversion. All of the calculations indicate that energetically the most favorable line of approach for the attacking hydrogen atom is along the axis of the H_2 molecule. It is therefore reasonable and convenient to concentrate attention on three-dimensional potential-energy surfaces in which the angle is held at 180° and in which potential energy is plotted against the $H^\alpha\!\!-\!\!H^\beta$ and $H^\beta\!\!-\!\!H^\alpha$ distances.

Two general quantum-mechanical procedures have been used in attacking this problem. The first is based on an application of an approximate quantum-mechanical equation proposed by F. London. This procedure was useful in the earlier days for arriving at the general form of the surfaces, but is not very accurate. Most modern work is therefore based on the variational method, and an account of this is given later.

Treatments based on the London equation The methods involving the London equation are briefly as follows. In the year following Schrödinger's first papers[1] on the new wave mechanics, Heitler and London[2] gave an approximate treatment of the energy of the H_2 molecule, using

[1] E. Schrödinger, *Ann. Physik.*, **79**:361, 489 (1926); **80**:437 (1926); **81**:109 (1926).
[2] H. Heitler and F. London, *Z. Physik*, **44**:455 (1927).

the wave functions

$$\psi = \psi_A(1)\psi_B(2) \pm \psi_B(1)\psi_A(2) \tag{1}$$

The subscripts A and B refer to the two nuclei, and 1 and 2 to the two electrons; the eigenfunction $\psi_A(1)$, for example, is the eigenfunction for the 1s state of electron 1 on nucleus A. The eigenvalues corresponding to (1) are

$$E = \frac{Q \mp J}{1 \mp S^2} \tag{2}$$

where Q, J, and S are the coulombic, exchange, and overlap integrals given by

$$Q = \int\!\!\!\int \psi_A(1)\psi_B(2) \mathsf{H} \psi_A(1)\psi_B(2)\ d\tau \tag{3}$$
$$J = \int\!\!\!\int \psi_A(1)\psi_B(2) \mathsf{H} \psi_B(1)\psi_A(2)\ d\tau \tag{4}$$
$$S = \int\!\!\!\int \psi_A(1)\psi_B(2)\psi_B(1)\psi_A(2)\ d\tau \tag{5}$$

H is the hamiltonian operator for the system. Each of these three integrals, and hence E, is a function of the intermolecular separation r. The integrals become zero at large separations, so that the energy $E(r)$ relates to an energy of zero for the completely separated atoms. The lower (positive) signs give rise to the lower energies, since the integrals are negative in value, and correspond to the bound state $^1\Sigma$ of the molecule. The upper, negative signs refer to the repulsive state. The Heitler-London equation accounts for 66 percent of the bonding in H_2; the experimental binding energy is 109.4 kcal and Eq. (2) leads to 72.4 kcal.

If the overlap integral S is ignored, Eq. (2) becomes

$$E = Q \mp J \tag{6}$$

This simple result that the energies of the $^1\Sigma$ and $^3\Sigma$ states are the sum and difference of the coulombic and exchange integrals is illustrated in Fig. 8. Paradoxically, the calculated energy at the minimum of the $^1\Sigma$ curve is now -107.5 kcal/mole, in much better agreement with the experimental value than that given by Eq. (2). There is, however, no justification for neglecting S, the better agreement being due to a cancellation of errors.

The extension of the Heitler-London equation to the three-atom system shown in Fig. 9 was considered by London.[1] If we imagine that the atom A is removed to infinity, there remains the diatomic molecule

---

[1] F. London, "Probleme der Modernen Physik," p. 104, Sommerfeld Festschrift, 1928; *Z. Elektrochem.*, **35**:552 (1929).

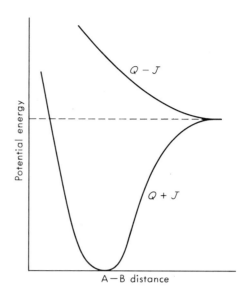

Fig. 8 Potential-energy curves for a diatomic molecule A—B, such as H_2. The curves correspond to the sum and difference of the coulombic integral Q and the exchange integral J.

B—C, the coulombic and exchange energies for which are designated as A and α. Similarly, B and β are the coulombic and exchange integrals for the isolated A—C molecule, and C and γ for the isolated A—B molecule. For the triatomic system shown in Fig. 9 London wrote down the equation

$$E = A + B + C \pm \{\tfrac{1}{2}[(\alpha - \beta)^2 + (\beta - \gamma)^2 + (\gamma - \alpha)^2]\}^{1/2} \quad (7)$$

This equation reduces to the form of Eq. (6) if any one of the atoms is removed to infinity; for example, if A is removed to infinity, the energies B, C, β, and γ all become zero, and Eq. (7) reduces to

$$E = A \pm \alpha \quad\quad\quad\quad\quad\quad\quad\quad\quad (8)$$

which is equivalent to (6). Equation (7) was given by London without

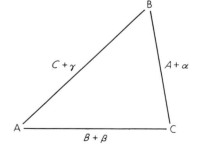

Fig. 9 A triatomic system. The terms A, B, and C represent the coulombic energies of the individual diatomic species, with the third atom moved to infinity; α, β, and γ are the corresponding exchange energies.

proof; Slater[1] gave a derivation of it, and this was amended by Coolidge and James.[2]   The derivations show that the equation can only be expected to apply to systems where the electrons are in s orbitals.

There are a number of reasons why the London equation cannot give very reliable energies.   In the first place, the Heitler-London equation gives for the H_2 binding energy an error which is much greater than the activation energy for the $H + H_2$ reaction.   The London equation is in fact related not to the original Heitler-London equation (2) but to the simplified form (6).   Admittedly, the latter happens to lead to a better result for H_2, but this is clearly fortuitous; we could not rely on a similar cancellation of errors for triatomic systems.

Aside from this point, the derivation by Coolidge and James of Eq. (7) shows that multiple-exchange integrals, i.e., terms arising from permutations of more than two electrons, are neglected; these neglected terms can be shown to be quite important.   Furthermore, the integrals A, B, C, α, β, and γ used in deriving the equations are assumed to be the same as if the third atom were removed; A and α, for example, are based on a hamiltonian operator for the B-C system instead of for the triatomic system.

In spite of these deficiencies the London equation has proved useful in giving the right general form for potential-energy surfaces.   In dealing with such surfaces one is concerned with the *difference* between the energy of the triatomic system and that of A + BC, and it is possible that by a cancellation of errors this difference would be fairly reliable.

The coulombic and exchange integrals Q and J for the isolated H_2 molecule were computed by Sugiura[3] as a function of interatomic distance. Eyring and Polanyi[4] considered the possibility of introducing his values into the London equation (7); they did not, however, perform the calculations because they thought that the errors would be too great.   The calculations have been made much more recently[5] and lead to the potential-energy surface shown in Fig. 10.   It is seen that the valleys do not meet at a single col, but that there is a potential-energy *basin* corresponding to a symmetrical H····H····H configuration.   The heights of the barriers on each side of the basin are seen to be 8.8 kcal/mole.   This is in surprisingly good agreement with the classical barrier height of 8 to 10 kcal deduced

[1] J. C. Slater, *Phys. Rev.*, **38**:1109 (1931).

[2] A. S. Coolidge and H. M. James, *J. Chem. Phys.*, **2**:811 (1934).

[3] Y. Sugiura, *Z. Physik*, **45**:484 (1927).

[4] H. Eyring and M. Polanyi, *Z. Physik. Chem. (Leipzig)*, **B12**:279 (1931).

[5] P. J. Kuntz, E. N. Nemeth, and J. C. Polanyi, private communication; see K. J. Laidler and J. C. Polanyi, in G. Porter (ed.), "Progress in Reaction Kinetics," vol. 3, Pergamon Press, Oxford, 1965.

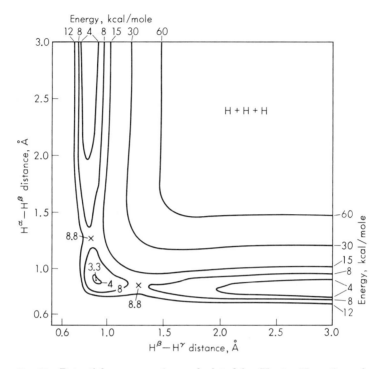

Fig. 10 Potential-energy surface calculated by Kuntz, Nemeth, and Polanyi for the linear H····H····H system. The calculation involves inserting Sugiura values for the coulombic and exchange integrals into the London equation. The energies, in kilocalories per mole, are expressed relative to that of $H + H_2$ (classical ground state).

from the experimental results.[1] However, this agreement must be regarded as fortuitous in view of the very serious approximations involved in the London equation. Moreover there is good reason to believe, on the basis of more reliable calculations to be considered later, that there is in fact no basin in the potential-energy surface for this reaction.

Some light on the approximations involved in the London treatment is provided by calculations of Coolidge and James[2] for the linear symmetrical H····H····H system in which the two bond distances are 0.9 Å. The binding energy relative to 3H, calculated by inserting Sugiura's

---

[1] As indicated in Chap. 1, there is no simple connection between the experimental activation energy and the barrier height, the relationship depending upon the details of how the process occurs. As a result, there is an uncertainty of at least 1 kcal in the barrier height that one can estimate from experiment.

[2] Coolidge and James, *loc. cit.*

values into the London equation, was 103.5 kcal; this is consistent with the values shown in Fig. 10. However, when they "improved" the London equation by including the overlap integral, taking into account multiple-exchange integrals, and using the complete hamiltonian operator, the calculated energy went up by nearly 30 kcal/mole. The treatment thus gives much worse agreement with experiment when these changes are made, and this shows that the agreement given by the simple Eq. (7) is fortuitous.

In spite of this, the early applications of the London equation, made largely by semiempirical methods to be considered later, were of great importance in that they provided a framework within which it was possible to understand the significance of the activation energy. The London equation in fact constituted the first quantum-mechanical statement of the idea that bond breaking is assisted by the formation of a new bond.

The London equation in its original form applies only to s orbitals. It can be modified to take account of p orbitals, and this was done by Magee[1] for reactions such as $Cl + H_2$ and $H + HCl$. His equation took account of double-exchange integrals, but not higher ones. The calculations yielded low energy barriers for these reactions and predicted a stable triangular complex.

Variational calculations   Much more reliable calculations of molecular energies are made on the basis of the variational principle, and most of the more recent work has been done in this way.

The hamiltonian operator H for any system can readily be written down, but the difficulty lies in the solution of the differential equation

$$H\psi = E\psi \tag{9}$$

The point of the variational principle is that it can be shown that if one selects *any* function φ, the energy given by the integral

$$\int \varphi H \bar{\varphi}\, d\tau$$

(where $\bar{\varphi}$ is the complex conjugate of φ) *cannot be below* the true energy E for the ground state of the system. We can thus calculate the integral for a variety of functions φ and know that the lowest energy obtained will be closest to the truth. More systematically, we can express φ as a function of one or more variables and calculate the integral as a function of these variables; the energy (i.e., the integral) can then be minimized with respect to the variables. The main difficulty with the method is that one does not know what type of function φ will be most satisfactory for a given problem.

[1] J. L. Magee, *J. Chem. Phys.*, **8**:677 (1940).

The first variational calculations of potential-energy surfaces were made by Hirschfelder, Eyring, and Rosen.[1]   Their results are summarized in columns 1 to 4 of Table 1.   The calculated activation energies relate to the classical ground states of the reactants and activated complexes and are to be compared with the value of 8 to 10 kcal/mole derived from the experimental results.   In this table, two calculated activation energies are given, E_c and E_c'.   The former is the energy-barrier height relative to the *experimental* value of the energy of H + H_2 (classical ground state), while E_c' is the barrier height relative to that for H + H_2 *calculated on the same basis*.   The calculations of E_c' compare more favorably with experiment because there will be some cancellation of errors when the same quantum-mechanical procedures are applied to H_3 as are applied to H + H_2.   However, the calculated value of E_c provides a more satisfactory test of the reliability of the calculations.

The procedures adopted by Hirschfelder et al. are as follows, with reference to columns 1 to 4 in Table 1:

1. Homopolar wave functions
2. The best linear combination of polar and homopolar wave functions
3. The best homopolar wave function with variable effective nuclear charge Z_{eff}, often referred to as a *screening* or *scale* factor
4. The best wave function with simultaneous variation of the polarity and the effective nuclear charge

In each case the wave functions were constructed from $1s$ hydrogen-like atomic orbitals, and only linear symmetrical configurations of H⋯H⋯H were considered.   Later Hirschfelder,[2] using an early computer, evaluated the three-center integrals for nonlinear symmetrical configurations; he showed that the linear configurations are the most stable, the equilateral triangle, for example, being about 80 kcal less stable than the linear system.

These calculations all lead to barrier energies that are considerably higher than the true ones; even the E_c' values are much higher.   It is to be noted that the inclusion of a variable Z_{eff} does not lead to much improvement (compare 3 with 1 and 4 with 2).   The reason may be that the same screening constant has been used for the central H atom as for the outer ones.   Barker and Eyring[3] independently varied the effective charge on the middle atom and that on the outer atoms and obtained an improvement of about 5 kcal (compare 4 and 5).

[1] J. O. Hirschfelder, H. Eyring, and N. Rosen, *J. Chem. Phys.*, **4**:121 (1936).

[2] J. O. Hirschfelder, *J. Chem. Phys.*, **6**:795 (1938); J. O. Hirschfelder and C. N. Weygandt, *J. Chem. Phys.*, **6**:806 (1938).

[3] R. S. Barker and H. Eyring, *J. Chem. Phys.*, **22**:1182 (1954).

Table 1 Variational calculations of the energy of the linear activated complex H_3 relative to $H + H_2$

	(1)	(2)	(3)	(4)	(5)	(6)	(7)	(8)	(9)	(10)	(11)	(12)
	Experi-mental[a] / Heitler-London	H-L plus polar	H-L plus Z_{eff}	H-L polar plus Z_{eff}	H-L polar plus two Z_{eff}	3-AOMOCI[b]	3-AOMOCI[b] plus Z_{eff}	5-AOMOCI[b] plus Z_{eff}	6-AOMOCI[b] plus Z_{eff}	27-AOMOCI[b]	6-AOMOCI[b]	15-AOMOCI[b]
Z, a.u.	1	1	1.06	1.09	$\{ Z_0 = 1.00,\ Z_m = 1.08$	1	1.10	1.10	$\{ Z_1 = 1.245,\ Z_4 = 0.83$	1	1	1
r, Å	1.06	1.06	1.00	0.975	0.95	~1.0	1.00_5	1.02	0.943	0.95	0.94	0.932
E_c, kcal/mole	56.3 {8-10}	49.0	53.2	42.3	~37	48.7	41.1	37.0	29.2	17.6	7.74	14.0
E'_c, kcal/mole	19.1 {8-10}	13.6	30.7	25.2	~20	13.3	24	19.9	15.4	14.3	7.74^c	11.0
	d	e	f	g	h	i	j	k	l	m	n	o

[a] There is considerable uncertainty about estimating the energy of the linear H_3 complex from experiment since the reaction does not necessarily proceed by a linear complex; see Chap. 7, especially pages 160 to 171.

[b] Atomic-orbital molecular-orbital-configuration interaction.

[c] This is not strictly a variational treatment and involves extrapolation; see text.

[d-g] J. O. Hirschfelder, H. Eyring, and N. Rosen, *J. Chem. Phys.*, **4**:121 (1936).

[h] R. S. Barker and H. Eyring, *J. Chem. Phys.*, **22**:1182 (1954).

[i] B. J. Ransil, *J. Chem. Phys.*, **26**:971 (1957).

[j] V. Griffing, J. L. Jackson, and B. J. Ransil, *J. Chem. Phys.*, **30**:1066 (1959).

[k] G. E. Kimball and J. G. Trulio, *J. Chem. Phys.*, **28**:493 (1958).

[l] S. F. Boys and I. Shavitt, *Univ. Wisc. Naval Res. Lab. Tech. Rep.*, Wis-AF-13 (1959).

[m] C. Edmiston and M. Krauss, *J. Chem. Phys.*, **92**:1119 (1965).

[n] H. Conroy and B. L. Bruner, *J. Chem. Phys.*, **47**:971 (1967).

[o] I. Shavitt, R. M. Stevens, F. L. Minn, and M. Karplus, *J. Chem. Phys.*, **48**:6 (1968).

For some 15 years after the early calculations were made little further work was done. Calculations during that period had to be done by hand or with very simple calculating machines or computers. More recently, with the development of high-speed computers, there has been a considerable revival of interest in the problem. Walsh and Matsen[1] calculated the binding energy of H_3, using several molecular-orbital approximations, but appear to have made an error in setting up their wave functions.[2] Their most complete treatment, which involved configuration interaction (CI),[3] was repeated by Ransil,[4] whose results are given in column 6 of Table 1. This molecular-orbital-configuration-interaction (MOCI) treatment is equivalent to the valence-bond treatment with polar structures included (column 2 of Table 1); the difference between 2 and 6 is indeed within the computational error. Griffing, Jackson, and Ransil[5] also carried out an MOCI treatment but allowed Z_{eff} to vary; their results are recorded in column 7 of Table 1, and as expected are very similar to the valence-bond treatment of column 4.

An important advance was made by Kimball and Trulio,[6] who, instead of constructing molecular orbitals from three atomic orbitals, used a larger number. For the linear symmetric H_3 complex they set up a molecular orbital as a linear combination of atomic orbitals (LCAOMO), using five $1s$ hydrogen-like orbitals, with centers equally spaced on the line of nuclei. Complete configuration interaction was included, and Z_{eff} was varied, being taken as the same for all five orbitals. The five orbitals could be moved freely relative to the nuclei, subject to the restriction of equal spacing and of symmetry. By varying Z_{eff}, the orbital positions, and the H—H distances, they obtained an energy at the activated state that was lower by 11.7 kcal than that given by Ransil's treatment involving three atomic orbitals (compare columns 6 and 8). The result is still, however, a good way from the experimental value. Kimball and Trulio attributed this rather disappointing outcome to the fact that even with three orbitals the charge distribution is still quite well adjusted along the molecular axis, so that the addition of the two orbitals in intermediate positions does not make a great difference. They suggested that, if they had not insisted on equal spacing of orbitals and had piled up charge

---

[1] J. M. Walsh and F. A. Matsen, *J. Chem. Phys.*, **10**:526 (1951).

[2] S. F. Boys and I. Shavitt, *Univ. of Wisc. Naval Res. Lab. Tech. Rept.*, WIS-AF-13 (1959).

[3] This is a procedure in which molecular orbitals are constructed by superimposing eigenfunctions of the same symmetry.

[4] B. J. Ransil, *J. Chem. Phys.*, **26**:971 (1957).

[5] V. Griffing, J. L. Jackson, and B. J. Ransil, *J. Chem. Phys.*, **30**:1066 (1959).

[6] G. E. Kimball and J. G. Trulio, *J. Chem. Phys.*, **28**:493 (1958).

on the middle nucleus by bringing the two extra orbitals toward the center atom, they would have achieved much better results.

A very complete variational treatment of H_3 has been carried out by Boys and Shavitt.[1] Instead of concentrating charge by the differential-screening method of Barker and Eyring or by the additional orbitals of Kimball and Trulio, they located *a pair* of $1s$ orbitals at each nucleus, with different effective nuclear charges for the members of each pair. Thus, at nucleus A there was a $1s$ orbital with $(Z_{eff})_1$ and another $1s$ orbital with $(Z_{eff})_2$; at B there was a $1s$ orbital with the same $(Z_{eff})_1$ as at A, and a $1s$ orbital with the same $(Z_{eff})_2$ as at A; similarly at C. The ratio $(Z_{eff})_1/(Z_{eff})_2$ was arbitrarily fixed at 1.5. It turned out that the energy of the system was not very sensitive to the value used for Z_{eff}, since the redistribution of charge took place "automatically" through variational optimization of the relative contributions from the two types of orbitals located at each nucleus; this feature probably obviates the need for different sets of Z_{eff} at each nucleus.

Calculations were made for bent and asymmetric configurations, but the linear configuration was found to be of lowest energy. In the symmetric activated complex the charge on the central atom proved to be about 10 percent greater than on the outer atoms. The binding energy, however, was still quite far from the experimental value (see column 9).

More recently Edmiston and Krauss[2] have published a brief report of calculations of the saddle-point energy of H_3, taken to be the energy where the H—H distances are 0.95 Å. Theirs was a configuration-inter-action study involving four s and five p gaussian orbitals on each of the three atoms. For the H_2 molecule the calculations led to a classical dissociation energy of 106.1 kcal, to be compared with the spectroscopic value of 109.4 kcal, and these figures led to the E_c and E_c' values listed in column 10 of Table 1. It is seen that as far as E_c is concerned there is a substantial improvement over the results of the previous calculations.

Conroy and Bruner[3] have carried out some very successful calcula-tions which, although not strictly based on the variation principle, are conveniently considered at this point. If one has obtained the exact eigenfunction for the operator H, the energy of the system is given exactly by the equation

$$E = \frac{H\psi}{\psi} \tag{10}$$

[1] S. F. Boys, G. B. Cook, C. M. Reeves, and I. Shavitt, *Nature*, **178**:1207 (1956); Boys and Shavitt, *op. cit.*

[2] C. Edmiston and M. Krauss, *J. Chem. Phys.*, **42**:1119 (1965).

[3] H. Conroy and B. L. Bruner, *J. Chem. Phys.*, **47**:971 (1967).

Conroy and Bruner proceeded by a series of successive approximations and extrapolations to obtain an eigenfunction that leads to consistent results at various positions relative to the molecule;[1] they then used Eq. (10) to calculate the energy. Since the variation principle is not involved, their eigenfunction can lead to energies that are lower than the true ones. They employed a molecular orbital made up of six atomic orbitals each one of which contained a considerable number (17 to 36) of variable coefficients. Their calculations led to exactly the right classical dissociation energy (109.4 kcal/mole) for the H_2 molecule, and their E_c and E'_c values are therefore the same; the value they obtained is 7.74 kcal, which one suspects is slightly below the experimental value (about which, as indicated earlier, there is some uncertainty). The complete potential-energy surface of Conroy and Bruner is shown in Fig. 11.

Up to the present time (1969) the most successful calculation based strictly on the variation principle is that of Shavitt, Stevens, Minn, and Karplus.[2] These workers made two sets of calculations, one based on six orbitals (two $1s$ orbitals on each nucleus) and one based on fifteen; the latter had two $1s$ orbitals and three $2p$ (i.e., $2p_x$, $2p_y$, and $2p_z$) orbitals on each nucleus. The 15-orbital treatment was more successful and led to the potential-energy surface shown in Fig. 12. The energy calculated for $H + H_2$ (involving a 10-AO treatment, with 5 AO's on each nucleus) was 3.0 kcal above the observed value, and the energy at the saddle point was 14.0 kcal above the observed value for $H + H_2$ (i.e., $E_c = 14.0$, $E'_c = 11.0$ kcal). Since the 10-AO calculation for H_2 is too high by 3.0 kcal, it might be reasonable to estimate that the calculation for H_3 will be $\frac{3}{2} \times 3.0 = 4.5$ kcal too high. This leads to $\sim$9.5 kcal for the barrier height, in good agreement with 9.13 estimated by a reliable semiempirical treatment (see page 164), but still significantly higher than the 7.7 kcal calculated by the extrapolation procedure of Conroy and Bruner. The dynamical calculations (see Chap. 7) suggest that the higher values are to be preferred, but the matter is by no means settled.

These 1968 calculations of Shavitt et al. lead to no basin at the saddle point. It appears that the more reliable quantum-mechanical and semiempirical calculations do not give a basin of any significant depth.

Although the $H + H_2$ system seems to have been dealt with fairly satisfactorily by this treatment, the situation regarding purely quantum-

---

[1] In practice they minimized the energy variance

$$u^2 = \frac{\int (H\psi - E\psi)^2 \, d\tau}{\int \psi^2 \, d\tau}$$

For details of the procedure used, see H. Conroy, *J. Chem. Phys.*, **41**:1336 (1964).

[2] I. Shavitt, R. M. Stevens, F. L. Minn, and M. Karplus, *J. Chem. Phys.*, **48**:6 (1968).

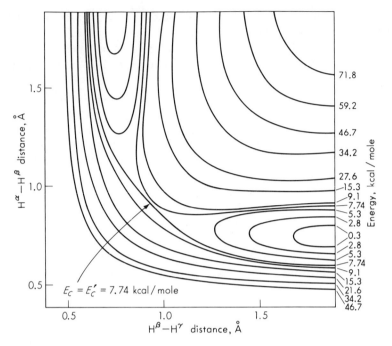

Fig. 11 Potential-energy surface for the H + H$_2$ reaction calculated by Conroy and Bruner using a quantum-mechanical procedure involving an extrapolation procedure. The calculated activation barrier (7.74 kcal) is consistent with the experimental value.

mechanical calculations still remains somewhat discouraging. An enormous amount of labor has gone into making the more reliable estimates for the energy of the activated complex, and the success has arisen less from the initial choice of good orbitals than from the mathematical device of varying a large number of parameters. We are thus not much further forward as far as the calculation of activation energies for more complex systems is concerned. It is likely that the amount of labor required to make reliable calculations for systems containing more than three electrons will be very great. It is always possible, of course, that a completely new computational approach will overcome this difficulty.

SEMIEMPIRICAL TREATMENTS

The calculations of potential-energy surfaces dealt with in the last section are purely quantum-mechanical ones; they of course make use of experimental quantities such as the electronic charge, but they do not use any of the quantities that they are designed to interpret, such as bond-dissociation energies.   No adjustments are made in these purely quantum-mechanical calculations with a view to obtaining a result that is closer to the experimental one.

By contrast, the semiempirical treatments, although based on quantum-mechanical theory, make use of experimental results of the kind they are interpreting, and adjustments are made to obtain more satisfactory results.   The extent to which this is done varies to a considerable extent; sometimes the adjustment is so extensive that the prefix "semi" seems to be an understatement.   There is no sharp distinction between a "semiempirical" and an "empirical" treatment, the latter leaning more heavily on the experimental results, and less (or hardly at all) on theory, than the former.

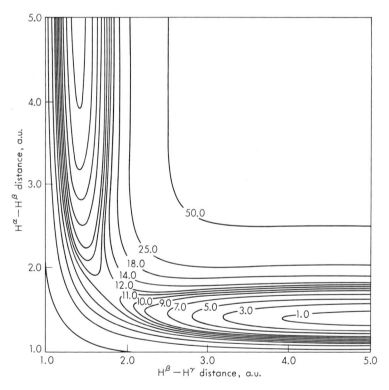

Fig. 12   The potential-energy surface calculated by Shavitt, Stevens, Minn, and Karplus, using a purely quantum-mechanical procedure based on the variation principle.   The barrier height is 11.0 kcal higher than the calculated H + H_2 level.

The London-Eyring-Polanyi method   The earliest of the theories usually referred to as semiempirical was due to Eyring and Polanyi[1] and was
based on the London equation (page 15); surfaces based on their method
are now usually known as *London-Eyring-Polanyi (LEP) surfaces*.   The
method involves neglecting the overlap integrals in the Heitler-London
and London equations, so that the energy for the ground state of the
diatomic molecule B—C in Fig. 9 (with A removed to infinity) is given
simply by

$$E = A + \alpha \tag{11}$$

The assumption is now made that the coulombic and exchange energies A
and α are constant fractions of the total energy for all internuclear
distances.   The total energy E can readily be obtained as a function of
internuclear distance r from analysis of spectroscopic data, and expressions
such as that of Morse have been of great value in relating the potential
energy to the distance; the Morse equation is

$$E = D(e^{-2\beta(r-r^0)} - 2e^{-\beta(r-r^0)}) \tag{12}$$

where r^0 is the equilibrium internuclear distance, D is the classical dissociation energy, and β is a constant.

From an inspection of Sugiura's calculations of the coulombic and
exchange integrals for the H_2 molecule, Eyring and Polanyi concluded
that over a range of interatomic distances (in particular, for $r > 0.8$ Å)
the fraction

$$\rho = \frac{A}{A + \alpha} \tag{13}$$

is roughly constant at 10 to 15 percent.   For any triatomic configuration
(see Fig. 9) it is therefore possible to evaluate for each pair of atoms
the coulombic and exchange energies, on the basis of the spectroscopic
value for the total energy; A, B, C, α, β, and γ can therefore readily
be calculated for the system, and insertion in the London equation (7)
then gives the required energy for the triatomic species.

Later workers have tried taking various percentages of coulombic
and exchange energies.   Sometimes the procedure has been to use the
percentage as an empirical parameter in order to obtain satisfactory
agreement for one particular reaction.   The activation barrier varies

---

[1] H. Eyring and M. Polanyi, *Z. Physik. Chem. (Leipzig)*, **B12**:279 (1931); for a transla-
tion into English of the first part of this article, see M. H. Back and K. J. Laidler,
"Selected Readings in Chemical Kinetics," p. 41, Pergamon Press, Oxford, 1967.
For a general review of the calculations, see S. Glasstone, K. J. Laidler, and H. Eyring,
"The Theory of Rate Processes," McGraw-Hill Book Company, New York, 1941.

rather strongly with the fraction ρ of coulombic energy; with $\rho = 20$ percent, the barrier height is 7 kcal; for $\rho = 14$ percent, it is 14 kcal; and for $\rho = 7$ percent, the value is 19 to 20 kcal.[1]

This method of Eyring and Polanyi has proved useful in making rough estimates of energies of activation, but is not capable of high accuracy. A reasonable value of ρ usually succeeds in accounting for the experimental barrier heights, but it is quite another thing to suggest that the method can be used for making predictions. To test its predictive value Eyring and coworkers in a series of papers calculated activation energies with ρ restricted to the range of 10 to 12 percent.[2] They also treated four-electron systems by an extension of the method. The results of their calculations are shown in column 1 of Table 2, where they are compared with experiment. There is seen to be a very rough correlation between the calculated and experimental values. Some of the agreement is undoubtedly fortuitous, since in all but two of the reactions p electrons are involved and for these the simple London equation should not be applicable.

It has been seen (Fig. 10) that insertion of Sugiura's values into the London equation gives rise to a surface that has a basin at the activated state. Figure 13 shows a potential-energy surface calculated by Eyring, Gershinowitz, and Sun[3] with ρ taken as 14 percent, and there is seen to be a basin some 2 to 3 kcal deep. The depth of the basin increases as ρ is increased.[4] For $\rho = 7$ percent there is no basin, but the activation barrier is too high (20 kcal); for $\rho = 20$ percent the basin is 5 to 6 kcal deep, and the barrier, 7 kcal, is now much closer to the experimental value. For the $H + H_2$ system the method cannot predict the right barrier height without giving a basin. It will be seen in Chap. 7 that the existence of a basin leads to the result that the activated complexes will have long lives, so that they can survive a few vibrations. The results of molecular-beam studies, however, have suggested that complexes of long life are unusual, so that some doubt is cast on the validity of these basins at the activated state. Also, as seen in the previous section, the most reliable quantum-mechanical calculations for this system do not show a basin.

[1] H. Eyring, H. Gershinowitz, and C. E. Sun, J. Chem. Phys., 3:786 (1935); J. C. Polanyi, J. Chem. Phys., 23:1505 (1955).

[2] See Glasstone, Laidler, and Eyring, op. cit.

[3] H. Eyring, H. Gershinowitz, and C. E. Sun, J. Chem. Phys., 3:786 (1935); J. O. Hirschfelder, H. Eyring, and B. Topley, J. Chem. Phys., 4:170 (1936).

[4] J. C. Polanyi, J. Chem. Phys., 23:1505 (1955); K. J. Laidler and J. C. Polanyi, in G. Porter (ed.), "Progress in Reaction Kinetics," vol. 3, Pergamon Press, Oxford, 1965.

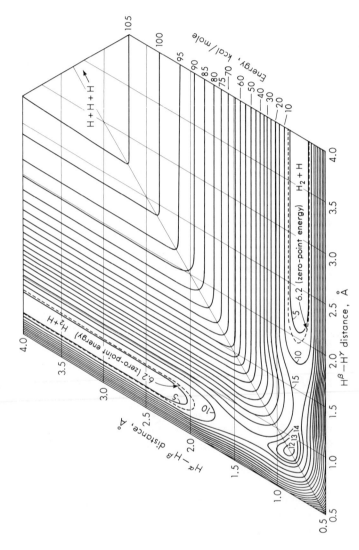

Fig. 13 Potential-energy surface for the H + H₂ reaction, calculated by Eyring, Gershinowitz, and Sun using the semiempirical LEP method with the coulombic energy taken to be 14 percent of the total energy. The axes are skewed because dynamical calculations (cf. Chap. 7) were made on the basis of this surface.

Table 2 Activation energies (in kilocalories per mole) calculated by various semiempirical and empirical methods

Reaction	E_a, experimental[a]	(1) Semiempirical[b]	(2) Sato[c]	(3) BEBO[d]	(4) Maximum force
F + H$_2$ → FH + H + 45	.....	10.6 (0.10) 6.3 (0.14)	0.5 (0.18)	2	10.4
H + Cl$_2$ → HCl + Cl + 45	2–3	2.7 (0.14)	....	..	0.7
H + Br$_2$ → HBr + Br + 41	0–1	3 (0.10) 2.1 (0.14)	....	..	0.3
H + I$_2$ → HI + I + 35	0	2.5 (0.10) 1.8 (0.14)	....	..	0.2
H + HI → H$_2$ + I + 32	1.5	10.6 (0.10) 7.7 (0.14)	0.3 (0.18)	1	
H + p-H$_2$ → o-H$_2$ + H	8	20 (0.7) 14 (0.14) 7 (0.20)	23.8 (0.0) 13.4 (0.1) 5.4 (0.18) 3.6 (0.2)	10	10.1
Cl + H$_2$ → ClH + H − 1	5.5	11.6 (0.20)	5.2 (0.18)	8	9.0
Br + H$_2$ → BrH + H − 16	18.5	25.5 (0.20)	17.1 (0.18)	21	23.2
I + H$_2$ → IH + H − 32	33.4	43.3 (0.10) 40.4 (0.14)	33 (0.18)	..	41.2
H$_2$ + I$_2$ → 2HI + 3	39–40	48.3 (0.14)	....	..	38.2

[a] For references to these approximate experimental activation energies, see K. J. Laidler and J. C. Polanyi, in G. Porter (ed.), "Progress in Reaction Kinetics," vol. 3, p. 14, Pergamon Press, Oxford, 1965.

[b] Fraction of coulombic energy is shown in parentheses.

[c] S^2 is shown in parentheses.

[d] Bond-energy–bond-order.

Another unsatisfactory feature of the LEP calculations is that they give rise to a very deep basin at much shorter interatomic distances.[1] This basin is only eliminated by taking large ρ values, for example, 30 percent. At $\rho = 12$ percent the basin for the H + H$_2$ system is about 75 kcal deep, and H$_3$ is predicted to be about 60 kcal stable with respect to H + H$_2$. Since more reliable calculations have failed to reveal any such stable complexes, it is safe to conclude that they represent a serious defect in the method of constructing surfaces.

The LEP procedure has been extended to four-atom systems; details are given by Glasstone et al.[2]

[1] S. Sato, Bull. Chem. Soc. Japan, **28**:450 (1955); J. Chem. Phys., **23**:592, 2465 (1955); I. Yasumori, Bull. Chem. Soc. Japan, **32**:1103, 1110 (1959).

[2] S. Glasstone, K. J. Laidler, and H. Eyring, "The Theory of Rate Processes," pp. 121–126, McGraw-Hill Book Company, New York, 1941.

Sato's method   Both of these potential-energy basins appear to arise from the assumption of constant coulombic and exchange fractions.   To eliminate the basin, Sato[1] has proposed an alternative method in which ρ is treated as a function of the internuclear separation r.   He obtained the dependence of ρ on r on the basis of the shape of the repulsive $^3\Sigma$ curve, which has been calculated a number of times for H_2 and can be regarded as well known.   According to the simplified Heitler-London treatment (with S neglected) the energy for the repulsive state of B—C is given by

$$E_r = Q - J \tag{14}$$

In order to have an analytic expression for E_r, Sato modified the Morse equation by changing the sign between the two exponential terms from minus to plus; he also divided the expression by 2, since he found that this improved the agreement with the experimental repulsive curve.   The resulting equation is

$$E_r = \frac{D}{2} \left(e^{-2\beta(r-r_0)} + 2e^{-\beta(r-r_0)} \right) \tag{15}$$

[cf. Eq. (12)].   The energy values given by this equation can then be set equal to $Q - J$; those for the ground state are set equal to $Q + J$, so that Q and J can be obtained from the two equations.   In this way Q and J values are calculated at a series of internuclear separations without any assumption of a constant ratio.

Instead of basing calculations for the triatomic system A-B-C on the original London equation (7), Sato introduced an overlap integral and used the form

$$E = \frac{1}{1 + S^2} \left(A + B + C \pm \left\{ \tfrac{1}{2}[(\alpha - \beta)^2 + (\beta - \gamma)^2 + (\gamma - \alpha)^2] \right\}^{1/2} \right) \tag{16}$$

Sato concluded on intuitive grounds that this equation would be valid if the overlap integrals S_A, S_B, and S_C for the three diatomic species are equal.   It certainly reduces to the correct Heitler-London forms [Eq. (2)] in that case, but there is no validity in the conclusion that a treatment of the kind given by Coolidge and James (see page 16) would lead to (16) if $S_A = S_B = S_C$.   Polanyi and coworkers[2] have investigated this point and find that (16) is only obtained if $S_C \ll S_A = S_B$ and if three-electron permutations are neglected.   It appears, therefore, that there is no theoretical basis for Eq. (16), which must be justified empirically.

The Sato method is just as arbitrary and empirical as the LEP method, and in a comparison of the methods Weston[3] has concluded

[1] S. Sato, *Bull. Chem. Soc. Japan,* **28**:450 (1955); *J. Chem. Phys.,* **23**:592, 2465 (1955); I. Yasumori, *Bull. Chem. Soc. Japan,* **32**:1103, 1110 (1959).

[2] E. M. Nemeth, J. C. Polanyi, S. D. Rosner, and C. E. Young, private communication.

[3] R. E. Weston, *J. Chem. Phys.,* **31**:892 (1959).

that there is not much to choose between them. Sato's method is preferable to the LEP method in that it leads to surfaces free of basins. It may be inferior, however, in giving barriers that are too thin. A consequence of this is that calculations for the Sato surface lead to considerably more tunneling than observed experimentally (see page 90). This argument is not, however, a completely convincing one since there is uncertainty in the theoretical treatment of tunneling itself.

As far as the prediction of activation energies is concerned the Sato method is roughly on a par with the LEP method, as is seen by comparing columns 1 and 2 of Table 2. Weston has pointed out that for $H + H_2$ Sato's treatment requires that $S^2 = 0.148$, whereas the true S^2 at the activated state is three times as great. Moreover, there is no justification for using a constant S^2 for all configurations; thus S^2 is roughly halved when the H—H distance is increased from the normal distance by 0.5 Å. Similarly, the use of the same S^2 in different reactions can only be justified empirically.

An extension of Sato's procedure has been made by Cashion and Herschbach,[1] who abandoned the Morse-like repulsive function [Eq. (15)] and replaced it by the results of recent calculations of the repulsive energies. With $S^2 = 0$ they calculated an activation energy of 8.9 kcal, in contrast to the value of 23.8 obtained by Sato on the same assumption. This illustrates the sensitivity of the calculated activation energy to the form of the repulsive curves. That the procedure of Cashion and Herschbach gave a better activation energy is not, however, of much significance, since S^2 is in fact not zero.

Modified LEP methods A few methods have been developed which are basically of the LEP type but avoid some of the approximations inherent in the London equation (see page 16). They are essentially LEP treatments in which some of the integrals are adjusted on the basis of empirical evidence.

Yasumori[2] used the effective charge Z_{eff} as an adjustable parameter and evaluated the coulombic and exchange integrals for the diatomic pairs. He selected the best Z_{eff} not by the variational method but by taking the value of Z_{eff} that gave the correct difference in energy between the bonding and antibonding states in the region of internuclear separation corresponding to the activated state. This procedure led to $Z_{eff} \approx 1.4$, which is considerably larger than that given by the variational method for H_2 ($Z_{eff} = 1.166$) or for H_3 ($Z_{eff} = 1.1$).[3] However, the electrons in

[1] J. K. Cashion and D. R. Herschbach, *J. Chem. Phys.*, **40**:2358 (1964).

[2] I. Yasumori, *Bull. Chem. Soc. Japan*, **32**:1103, 1110 (1959).

[3] S. C. Wang, *Phys. Rev.*, **31**:579 (1928).

H_3 are simultaneously attracted to three nuclei, so that this larger Z_{eff} may be physically reasonable. Yasumori's potential-energy surface, calculated from the London equation, was free of basins and had a barrier height of 8.5 kcal, in good agreement with experiment. The calculated H—H distance was 0.88 Å.

Porter and Karplus[1] have also calculated a potential-energy surface on the basis of a modified LEP treatment. They calculated some of the energy contributions theoretically and others semiempirically. As in the London equation, they expressed the coulombic binding as the sum of the three contributions for the diatomic pairs. The exchange integrals were treated properly as involving the three-atom hamiltonian. Each exchange integral J was decomposed into a diatomic exchange integral J_d and a residual term Δ. The diatomic coulombic and exchange terms Q_d and J_d were obtained by Cashion and Herschbach's modification of Sato's method. The overlap integrals were calculated from $1s$ orbitals with approximately the same Z_{eff} as used by Wang[2] in his calculation on H_2. The residual terms Δ were calculated by using $1s$ wave functions, with $Z_{eff} = 1$, and by using an adjustable multiplier which is the ratio of the empirical J_d to the Heitler-London J_d, namely, 1.12. The double-exchange integrals were included; these were taken as proportional to the product of the three overlap integrals, $\varepsilon S_1 S_2 S_3$, where ε was calculated to be 0.616 by using $1s$ integrals.

Potential-energy surfaces were calculated in this way for a variety of H$\cdots$$\hat{\text{H}}$$\cdots$H angles. The results for the linear system and for an angle of 135° are shown in Fig. 14. There are no basins, and the classical barrier height for the linear complex is 8.58 kcal, corresponding to H—H distances of 0.896 Å. The calculations showed that the properties in the saddle-point region are very sensitive to the presence of the correction term Δ and of the double-exchange integrals.

From the theoretical standpoint Porter and Karplus's potential-energy surface is probably the most satisfactory semiempirical one. It is still impossible to say how closely it corresponds to the truth.

The bond-energy–bond-order method   Johnston and coworkers[3] have developed a bond-energy–bond-order (BEBO) treatment which is very

---

[1] R. N. Porter and M. Karplus, *J. Chem. Phys.*, **44**:1105 (1964); see also L. Pederson and R. N. Porter, *J. Chem. Phys.*, **47**:4751 (1967).

[2] Wang, *loc. cit.*

[3] H. S. Johnston, *Advan. Chem. Phys.*, **3**:131 (1960); H. S. Johnston and P. Goldfinger, *J. Chem. Phys.*, **37**:700 (1962); H. S. Johnston and C. Parr, *J. Am. Chem. Soc.*, **85**:2544 (1963); also H. S. Johnston, "Gas Phase Reaction Rate Theory," pp. 177–183, The Ronald Press Company, New York, 1966.

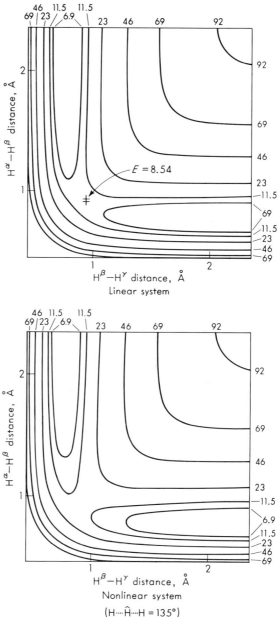

Fig. 14 Potential-energy surfaces for the $H + H_2$ reaction, as calculated by Porter and Karplus using a modified LEP procedure. The energies shown are in kilocalories.

successful in giving activation energies in good agreement with experiment (see column 3 of Table 2). It has no adjustable parameters but involves a considerable amount of empiricism; however, it seems more satisfactory to classify it as semiempirical than as empirical since the empirical relations it employs lie outside the field of kinetics.

One of the empirical relations it uses is an equation proposed by Pauling to relate bond length r and bond order n:

$$r = r_s - 0.26 \ln n \tag{17}$$

where r_s is the single-bond length for a bond connecting two particular atoms. Bond energy D can also be related to single-bond energy D_s through the relationship

$$D = D_s n^p \tag{18}$$

where p is a constant. Johnston and Parr calculated the exponent from the equilibrium internuclear separation and bond energy in a noble-gas diatomic cluster ($n \approx 0$).

Johnston and coworkers postulate that *for hydrogen-atom transfer* the path of lowest energy from reactants to products is defined by

$$n_1 + n_2 = 1 \tag{19}$$

where n_1 and n_2 are the orders of the bonds being broken and formed. Through (17) this controls the changes in bond lengths, and through (18), the changes in energy of the A—B and B—C bonds. This involves the assumption that the bonds can be treated separately. A correction is made for A-C repulsion with the use of Sato's modified Morse function for the repulsive state.

This method has been very successful in arriving at reliable estimates of activation energies, as is shown not only by the values in Table 2 but by the results of several other sets of calculations.[1]

COMPARISON OF SURFACES

Besides leading to different estimates of the activation barrier, the methods for calculating potential-energy surfaces lead to various features which are significant for interpreting reaction dynamics. The most important of these features are the presence or absence of a basin and the thickness of the barrier. The former feature is of great significance

[1] S. W. Mayer, L. Schieler, and H. S. Johnston, *J. Chem. Phys.*, **45**:385 (1966); also in *Proceedings of the Eleventh International Symposium on Combustion*, p. 837, The Combustion Institute, Pittsburgh, Pa., 1967; S. W. Mayer, *J. Phys. Chem.*, **71**:4159 (1967); S. W. Mayer and L. Schieler, *J. Phys. Chem.*, **72**:236 (1968).

in determining whether a complex of long life will be formed.   The thickness of the barrier influences the extent of quantum-mechanical tunneling.

The various calculations that have been made of potential-energy surfaces have led to quite different conclusions in this matter.   Some of these conclusions are demonstrated in Fig. 15, which shows energy profiles along the minimum-energy reaction path, for the linear H····H····H surfaces.   The original calculations of Eyring and Polanyi, and later similar calculations by the semiempirical method based on the London equation, led to a very broad and pronounced basin (curve E). · However, the more recent semiempirical calculations (curves S and PK) show no minimum. Most of the purely quantum-mechanical calculations (e.g., curve SSMK) confirm the absence of a basin.   The calculations of Conroy and Bruner (curve CB), however, were done at rather widely spaced distances, so that it is hard to judge if a basin exists; if it does, it is certainly not as broad and deep as suggested by the earlier calculations.

The evidence at the present time thus suggests that for the $H + H_2$ system there is either no basin or a very small basin at the saddle point. It certainly does not appear that the basin is substantial enough to have

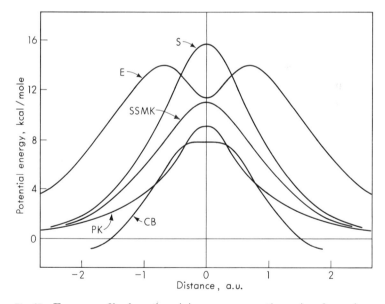

Fig. 15   Energy profile along the minimum-energy path, as given by various computations: E, early semiempirical calculations by Eyring and coworkers; S, Sato's semiempirical method; PK, refined semiempirical procedure of Porter and Karplus; CB, quantum-mechanical extrapolation procedure of Conroy and Bruner; SSMK, variational calculations by Shavitt, Stevens, Minn, and Karplus.

a significant effect on the results of dynamical calculations. For certain other reactions, however, there is both experimental and theoretical evidence suggesting the existence of a sufficiently deep basin to give rise to a complex of fairly long life and to affect the overall kinetic behavior.

As to the general shape and width of the barrier, there is fairly good agreement between the Sato, the Porter and Karplus, and the Shavitt, Stevens, Minn, and Karplus treatments. The earlier semiempirical treatments led to barriers that were substantially broader. Unfortunately, both the theoretical and experimental situations regarding tunneling are such that it is not easy to decide whether the broader or narrower barriers are more realistic.

EMPIRICAL TREATMENTS

A number of empirical relationships have been proposed which relate activation energy to other quantities, such as heats of reaction. Such relationships are useful in allowing rough estimates of activation energies to be made quite easily. In addition, a few empirical procedures have been given for constructing complete potential-energy surfaces.

Empirical activation energies   Evans and Polanyi[1] proposed a linear relationship between activation energies E and heats of reaction, ΔH, (or heats evolved, $Q = -\Delta H$) and discussed it in terms of potential-energy profiles; their equation can be written as

$$E = \alpha \, \Delta H + c = -\alpha Q + c \tag{20}$$

where the constant α is between zero and unity. The significance of this equation may be understood with reference to the potential-energy profiles shown in Fig. 16. Curve I relates to the system A + BC and shows the variation of potential energy with B—C distance, with the A—B distance held at the distance in the activated state. Curve II shows the corresponding variation for the system AB + C. Curve I' is part of the potential-energy curve for BC when A is a significant distance away, and curve II' is the curve for AB + C with the A—B distance corresponding to the normal molecule. The activation energy E is shown on the diagram; in fact there will be some resonance splitting at the crossing point of curves I and II. The heat evolved in the reaction, $Q \; (= -\Delta H)$, is also shown.

The dashed curves III and III' show the changes that occur if the

[1] M. G. Evans and M. Polanyi, *Trans. Faraday Soc.*, **34**:11 (1938); cf. C. N. Hinshelwood, K. J. Laidler, and E. W. Timm, *J. Chem. Soc.*, no. 848 (1938); S. Glasstone, K. J. Laidler, and H. Eyring, "The Theory of Rate Processes," pp. 139–146, McGraw-Hill Book Company, New York, 1941.

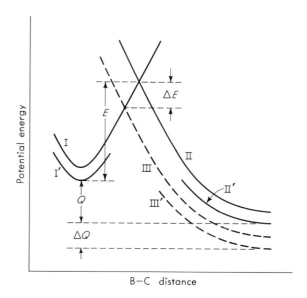

Fig. 16 Potential-energy profiles for reaction $A + BC \rightarrow$ $AB + C$, showing the relationship between the decrease ΔE in energy of activation and the increase ΔQ in heat evolved.

reacting species A is replaced by something else. Curve I is hardly affected, but if the A—B bond is strengthened by the change, the curve is lowered and the heat evolved increases by an amount ΔQ. In ideal cases the change in E will closely parallel that in Q. If curves I and II are symmetrical where they intersect, the decrease in E will be one-half the increase in Q or one-half the decrease in ΔH (i.e., $\alpha = 0.5$). Similar arguments can be used to interpret the changes in activation energy that occur when B or C is changed.

An extension of the Evans-Polanyi equation, which includes terms for both bond forming and bond breaking, has been proposed by Szabó;[1] his equation is

$$\Delta E = \sum_i D_i \text{ (breaking)} - \alpha \sum_j D_j \text{ (forming)} \tag{21}$$

where D_i and D_j are bond dissociation energies. Szabó has applied his relationship to several series of homologous reactions and finds it to be obeyed somewhat better than the simple Evans-Polanyi equation.

[1] Z. G. Szabó, *Chem. Soc. (London), Spec. Publ.*, **16**:113 (1962); Z. G. Szabó and T. Bérces, *Z. Phys. Chem.*, **57**:3 (1968).

Similar types of relationships have been developed by Bagdasaryan[1] and by Tikhomirova and Voevodsky,[2] and have been generalized by Semenov.[3] For a wide variety of abstraction reactions Semenov plots the energy of activation, E, against the heat evolved Q and finds that for exothermic reactions (Q positive) almost all the points lie fairly close to the line represented by the equation

$$E = 11.5 - 0.25Q \tag{22}$$

As shown in Fig. 17, the activation energies rarely deviate from this relationship by more than 1 kcal. For endothermic reactions the best procedure is to estimate the energy of activation of the reverse reaction

[1] K. S. Bagdasaryan, *Zh. Fiz. Khim.*, **23**:1375 (1949).
[2] N. N. Tikhomirova and V. V. Voevodsky, *Dokl. Acad. Nauk SSSR*, **79**:993 (1951).
[3] N. N. Semenov, "Some Problems in Chemical Kinetics and Reactivity," chap. 1, Pergamon Press, London, and Princeton University Press, Princeton, N.J., 1958.

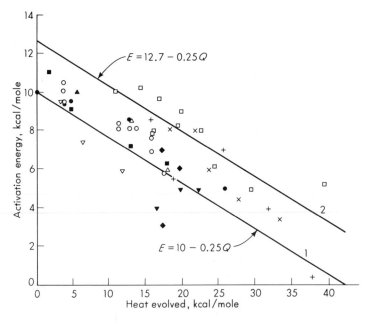

Fig. 17 Plot of activation energy against heat evolved, for a number of exothermic reactions, as follows:

● H + RH → H₂ + R	× H + RCl → HCl + R
○ CH₃ + RH → CH₄ + R	▼ CH₃ + RCl → CH₃Cl + R
■ D + RH → HD + R	◆ H + RBr → HBr + R
+ OH + RH → H₂O + R	▲ CH₃ + RBr → CH₃Br + R
△ H + RCHO → H₂ + RCO	□ Na + RCl → NaCl + R

by using Eq. (1) and then to calculate that for the forward reaction; this is equivalent to saying that for a negative value of Q the relationship is

$$E = 11.5 + 0.75Q \qquad (23)$$

These formulations are to be preferred to an early suggestion by Hirschfelder[1] that for an exothermic reaction of the type

$$A + BC \rightarrow AB + C$$

the activation energy is given by

$$E = 0.055D_{B-C} \qquad (24)$$

where D_{B-C} is the dissociation energy of the bond that is being broken. This equation implies that E is independent of the reactant A, which is only very roughly true.

It is to be noted that relationships such as those mentioned above are only reliable for essentially homolytic mechanisms. They are apt to break down for reactions in which there is considerable charge development in the activated state.[2]

Another type of empirical procedure, based on theoretical considerations, has been proposed by Otozai.[3] The theoretical basis of the method is, however, somewhat obscure, in contrast to the method of Johnston which has been classed as semiempirical; it seems better to class Otozai's method as empirical. Otozai assumed that the bond distances in the activated complex are equal to the bond distances which correspond to the maximum attractive *force* between the pair of molecules in question (treated as an isolated diatomic molecule). This point of maximum attractive force is the steepest point on the attractive side of the potential-energy curve. Otozai obtained $r^{\ddagger}$ by differentiating the equation

$$E = D[a(r - r^0) + 1]e^{-a(r-r_0)} \qquad (25)$$

This equation is somewhat similar to the Morse equation (12) and was given by Rydberg.[4] Differentiation of (25) leads to

$$r = r^0 + \frac{1}{a} \approx r^0 + 0.37 \text{ Å} \qquad (26)$$

for the distance where the force is a maximum; i.e., the bond has been extended by about 0.37 Å from its normal distance.

For a reaction of the type A + BC → AB + C the estimated activa-

[1] J. O. Hirschfelder, *J. Chem. Phys.*, **9**:645 (1941).

[2] Cf. A. Maccoll, *Chem. Soc. (London), Spec. Publ.*, **16**:116 (1962).

[3] K. Otozai, *Bull. Chem. Soc. Japan*, **24**:218 (1951).

[4] R. Rydberg, *Z. Physik*, **73**:376 (1932); **80**:514 (1933).

tion energy was derived from the A—B and B—C bond energies corresponding to these bond extensions in the activated state. In addition, an allowance was made for the energy of "activated bonds," which are the "bonds" created in the activated state, the A—C bond in the present case. To arrive at this energy, Otozai postulated that the sum of the energy of the activated bonds must be minimized, which in the present case means that the complex A—B—C must be linear. Some predictions given by this maximum force method are given in column 4 of Table 2.

Empirical potential-energy surfaces   A number of suggestions have been made for the construction of complete potential-energy surfaces on a purely empirical basis. Two types of approach have been made to this problem. In the methods of the first type, the Morse function has been adapted to give a surface; this is a reasonable procedure since it ensures that a reliable potential-energy curve will be obtained if for the A + BC system one of the atoms is removed to infinity. This method was employed by Wall and Porter,[1] who generated a surface for a linear triatomic system by "rotating" a Morse curve through 90° about a fixed point corresponding to large A—B and B—C distances. As it rotated, the Morse curve was systematically distorted, undergoing a general rise and fall so as to sweep out the two valleys and lead to a saddle point. This procedure is particularly convenient for symmetric reactions.

A somewhat similar procedure was employed by Blais and Bunker,[2] who distorted the Morse function of B—C by reducing its depth as A approached. This procedure is more convenient for nonlinear and asymmetric systems. Karplus and Raff[3] have improved this procedure by introducing a term which allows for some repulsion between A and BC.

The second type of approach to empirical potential-energy surfaces is to make modifications to the semiempirical surfaces, such as the LEP and Sato surfaces. Thus Polanyi[4] has suggested a modified Sato surface in which three overlap terms are added in a somewhat arbitrary way.

[1] F. T. Wall and R. N. Porter, *J. Chem. Phys.*, **36**:3256 (1962).
[2] N. C. Blais and D. L. Bunker, *J. Chem. Phys.*, **37**:2713 (1962); **39**:315 (1963).
[3] M. Karplus and L. M. Raff, *J. Chem. Phys.*, **41**:1267 (1964).
[4] J. C. Polanyi, *J. Quant. Spectr. Radiative Transfer*, **3**:471 (1963).

3
Activated-complex Theory

A potential-energy surface was first constructed in 1931 by Eyring and Polanyi,[1] and for a few years after that time a number of attempts were made to treat the dynamics of classical motion over potential-energy surfaces. The original paper by Eyring and Polanyi did in fact include a dynamical treatment of the $H + H_2$ reaction, and Eyring and his coworkers continued to pursue this problem for a number of years. There is no difficulty in principle in carrying out such dynamical calculations, but there was a practical difficulty. To cover a variety of initial conditions and obtain results that are statistically meaningful, it is necessary to carry out a very large number of calculations. In these earlier years high-speed computers had not been developed; consequently, even a great deal of labor led only to a limited number of results, which were of no statistical significance. The development of this very important field, which is now generally known as *molecular dynamics*, therefore had to await the development of high-speed computers. An account of the more recent work of this kind is given in Chap. 7.

[1] H. Eyring and M. Polanyi, *Z. Physik. Chem. (Leipzig),* **B12**:279 (1931).

In the meantime a number of other approaches to rate theory were being made. Prior to the development of the concept of potential-energy surfaces, rates had been treated by a theory of collisions in which molecules were treated as hard spheres. This theory was due to Trautz[1] and to W. C. McC. Lewis.[2] Such methods had some usefulness, but were altogether too crude and led to considerable error for many types of reactions. There was great need for a theory which was better than the simple collision theory and at the same time much easier to handle than the dynamical treatments, which inevitably require very extensive computations.

It was found possible to formulate a very satisfactory theory by focusing attention on the activated complexes, which are the molecular systems whose configurations correspond to the saddle-point region on the potential-energy surface. In particular, this theory arrives at an expression for the concentration of activated complexes through an equilibrium formulation; in a special sense, to be discussed in the next section, activated complexes are regarded as being at equilibrium with the reactant molecules. In this book the theory of rates that focuses attention on activated complexes, and calculates their concentrations on the basis of the equilibrium hypothesis, will be referred to as *activated-complex theory*. Various other expressions are used, including "absolute-rate theory," "the theory of absolute reaction rates," and "transition-state theory." The term *absolute* seems inappropriate in view of the present state of our knowledge of rate theory, and the term *transition state* is ambiguous since it is often applied to complexes other than activated complexes. The term *activated-complex theory* seems to be free of objections of this kind.

The activated-complex theory is based on an early suggestion by Marcelin,[3] who considered that the reacting molecules cross a "critical surface in phase space," which corresponds to the crossing of the potential-energy barrier. His idea was somewhat further developed by Rodebush[4] and later by O. K. Rice and Gershinowitz.[5] Pelzer and Wigner[6] carried out calculations, for reactions between hydrogen atoms and various mole-

[1] M. Trautz, *Z. Anorg. Allgem. Chem.*, **96**:1 (1916).

[2] W. C. McC. Lewis, *J. Chem. Soc.*, **113**:471 (1918).

[3] A. Marcelin, *Ann. Phys. (Paris)*, **3**:158 (1915); cf. A. March, *Z. Physik.*, **18**:53 (1917); R. C. Tolman, *J. Am. Chem. Soc.*, **42**:2506 (1920); **44**:75 (1922); E. P. Adams, *J. Am. Chem. Soc.*, **43**:1251 (1921).

[4] W. H. Rodebush, *J. Am. Chem. Soc.*, **45**:606 (1923); *J. Chem. Phys.*, **1**:440 (1933); **3**:242 (1935); **4**:744 (1936).

[5] O. K. Rice and H. Gershinowitz, *J. Chem. Phys.*, **2**:853 (1934); **3**:479 (1935); G. B. Kistiakowsky and J. R. Lacher, *J. Am. Chem. Soc.*, **58**:123 (1936).

[6] H. Pelzer and E. Wigner, *Z. Physik. Chem.*, **B15**:445 (1932).

cules, on the basis of this concept.   A particularly clear formulation of the situation was made in 1935 by Eyring,[1] and a somewhat similar approach leading to the same rate equation was made by Evans and Polanyi.[2]

The basic rate equation derived by Eyring, and by Evans and Polanyi, has been applied to a vast number of chemical reactions of all types and to a number of physical processes such as viscous flow.   By and large, it has proved to give a very satisfactory interpretation of reaction rates.   The great virtue of the theory is that it leads to so much at the cost of relatively little effort.   When attempts are made to improve the treatment, the "law of diminishing returns" sets in; a vast amount of computational effort is required, and sometimes very little or no improvement is achieved.   The situation at the present time (1969) is indeed that rates cannot be calculated any more satisfactorily by the more detailed theories than by activated-complex theory.   The more detailed theories do, however, provide information about certain aspects of rates, such as energy distribution in the product molecules, which is not dealt with at all by activated-complex theory.

In the present chapter the assumptions of activated-complex theory are discussed in some detail, and the basic equation is derived from several different points of view.

THE EQUILIBRIUM HYPOTHESIS

Consider the reaction

$$A + B \rightleftharpoons C + D$$

which has proceeded to equilibrium.   Under these conditions the activated complexes $X^{\ddagger}$ will also be in equilibrium with the reactants and products, and their concentration may therefore be calculated accurately by the methods of statistical mechanics in terms of the concentrations of A and B.   Under these conditions activated-complex theory does give a reliable treatment of the rate of the reaction; the theory therefore does apply accurately to the equal and opposite rates for a system at equilibrium.

The theory involves the hypothesis, however, that, even when the reactants and products are not at equilibrium with each other, the activated complexes are at equilibrium with the reactants.   Some justification for this assumption is provided in the following way.   Figure 18 shows a schematic energy diagram for a reaction system, and it will first

[1] H. Eyring, *J. Chem. Phys.*, **3**:107 (1935); W. F. K. Wynne-Jones and H. Eyring, *J. Chem. Phys.*, **3**:492 (1935).

[2] M. G. Evans and M. Polanyi, *Trans. Faraday Soc.*, **31**:875 (1935); **33**:448 (1937) M. Polanyi, *J. Chem. Soc.*, no. 629 (1937).

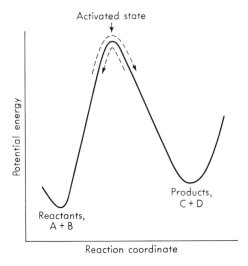

Fig. 18 A schematic energy diagram for the system $A + B \rightleftharpoons C + D$, showing the flow of activated complexes in the two directions.

be supposed that the reactants and products are at equilibrium with each other. At any instant of time there will be in the reaction vessel a few activated complexes, some of which are crossing the free-energy barrier in one direction, and some in the other. If the immediate past history of each of these activated complexes is investigated, some will be found to have been reactant molecules, and some product molecules. At equilibrium there will be an equal number of complexes of the two kinds, since the reaction is occurring at equal rates in the two directions.

Suppose that the product molecules are suddenly removed from the reaction system. The flow of those activated complexes that began as products (i.e., the flow from right to left in Fig. 18) will at once cease. There will still be a flow from left to right, however, and the assumption is that the rate of flow in this direction is unaffected by the removal of the products; in other words, the fluxes in the two directions are assumed to be independent of each other. It should be emphasized that, when one states that the activated complexes are in equilibrium with the reactants, *one is referring only to those complexes that, in the immediate past, were reactant molecules.*

There is no assumption that there is a classical type of equilibrium between initial and activated states; addition to the system of activated complexes moving from the initial to the final state would not disturb the equilibrium, as would be required if the equilibrium were classical. The activated complexes are transient species, passing from the initial to the final state and unable to turn back. They do not linger at the activated state and make a "decision" on whether to go forward or back; those that started as reactants are bound to go forward. This type of process that is envisaged by activated-complex theory is now often referred to as *direct* reaction (see also page 169); it is to be contrasted with an *indirect* reaction (sometimes known as *complex* reaction), in which the intermediate complexes are of long life and do sometimes go back rather than forward. The important point to note is that even in a direct reaction the complexes are, to a good approximation, formed in equilibrium states in the sense explained above.

Two other considerations also support the assumption of equilibrium. As noted above, the equilibrium assumption is certainly valid when reactants and products are at equilibrium, since then all species are at equilibrium; the theory therefore correctly predicts rates at equilibrium. If the theory were significantly in error before the establishment of equilibrium, rate constants would be expected to change as equilibrium is approached. Such a phenomenon has not been observed.

Secondly, a few very detailed calculations of rate constants have been made with methods that do not make the assumption of equilibrium; some of these involve the use of nonequilibrium statistical mechanics, and some are "stochastic" methods. Several such treatments are referred to in Chap. 8. Here it may simply be noted that these calculations lead to the conclusion that reactions are satisfactorily interpreted on the equilibrium assumption provided that E/RT for the reaction has a value of 5 or larger. The significance of this is that, if E/RT has a value smaller than 5, the reaction will occur so rapidly that there can no longer be equilibrium *even among the reactant molecules;* the more energetic species will be removed more rapidly than the supply of them can be replenished, and there will therefore not be a Boltzmann distribution of reactant molecules. Aside from this situation, however, the equilibrium theories lead to no significant error.

DERIVATIONS OF THE RATE EQUATIONS

Three derivations will be given of the rate equation based on activated-complex theory. These three derivations have much in common with one another—indeed, in a sense they are equivalent—but a careful consideration and comparison of the three derivations are of great help in understanding some of the rather subtle points that lie behind the treatment.

Derivation I In the first derivation to be given, the motion of the activated complexes over the top of the energy barrier is treated as a *very loose vibration.*

The equilibrium between reactants A + B and activated complexes

$X^{\ddagger}$ may be expressed by the equation

$$\frac{[X^{\ddagger}]}{[A][B]} = K^{\ddagger} \tag{1}$$

where $K^{\ddagger}$ is the equilibrium constant. This equilibrium constant may be written in terms of the appropriate partition functions:

$$\frac{[X^{\ddagger}]}{[A][B]} = \frac{Q^{\ddagger}}{Q_A Q_B} e^{-E_0/RT} \tag{2}$$

where E_0 is the difference between the zero-point energy per mole of the activated complexes and that of the reactants. Since this energy is the amount of energy that the reactants must acquire at $0°K$ before they can react, E_0 is the hypothetical energy of activation at this temperature. The partition functions in this expression must be evaluated with respect to the zero-point levels of the respective molecules.

These functions can be factorized into contributions corresponding to translational, rotational, vibrational, and electronic energy. The corresponding partition functions are given in Table 3. If, for example, the molecule A consists of N_A atoms, there will be $3N_A$ such partition functions, of which three are for translational motion, three for rotational motion (or two if the molecule is linear), and therefore $3N_A - 6$ for vibrational motion ($3N_A - 5$ for a linear molecule). The same is true for the activated complex, which consists of $N_A + N_B$ atoms, giving $3(N_A + N_B) - 6$ vibrational terms if the molecule is nonlinear. One of these vibrational factors is of a different character from the rest, since it corresponds to a very loose vibration which allows the complex to dissociate into the products C and D. For this one degree of freedom one may therefore employ, in place of the ordinary factor $(1 - e^{-h\nu/kT})^{-1}$, the value of this function calculated in the limit at which ν tends to zero. This is evaluated by expanding the exponential and taking only the first term

$$\lim_{\nu \to 0} \frac{1}{1 - e^{-h\nu/kT}} = \frac{1}{1 - (1 - h\nu/kT)} = \frac{kT}{h\nu} \tag{3}$$

The equilibrium constant may therefore be expressed by including this term $kT/h\nu$ and replacing $Q^{\ddagger}$ with $Q_{\ddagger}$, which now refers only to $3(N_A + N_B) - 7$ degrees of vibrational freedom [$3(N_A + N_B) - 6$ for a linear complex]; the resulting expression is

$$\frac{[X^{\ddagger}]}{[A][B]} = \frac{Q_{\ddagger}(kT/h\nu)}{Q_A Q_B} e^{-E_0/RT} \tag{4}$$

This expression rearranges to

$$\nu[X^{\ddagger}] = [A][B] \frac{kT}{h} \frac{Q_{\ddagger}}{Q_A Q_B} e^{-E_0/RT} \tag{5}$$

The frequency ν is the frequency of vibration of the activated complexes in the degree of freedom corresponding to their decomposition; it is therefore the frequency of decomposition. The expression on the left-hand side of Eq. (5) is therefore the product of the concentration of complexes $X^{\ddagger}$ and the frequency of their decomposition; it is therefore the rate of reaction, which is thus given by the expression on the right-hand side of the equation, i.e.,

$$v = [A][B] \frac{\mathbf{k}T}{h} \frac{Q_{\ddagger}}{Q_A Q_B} e^{-E_0/RT} \tag{6}$$

The result that the rate is proportional to the product [A][B] arose from the assumption that the activated complex is composed of one molecule of A and one of B; in the more general case of a complex composed of a molecules of A, b of B, and so on, the rate would be proportional to

Table 3 Partition functions

Motion	Degrees of freedom	Partition function	Order of magnitude
Translational	3	$\dfrac{(2\pi m \mathbf{k}T)^{3/2}}{h^2}$ (per unit volume)	10^{24}–10^{25}
Rotational† (linear molecule)	2	$\dfrac{8\pi^2 I \mathbf{k}T}{h^2}$	10–10^2
Rotational† (nonlinear molecule)	3	$\dfrac{8\pi^2(8\pi^3 ABC)^{1/2}(\mathbf{k}T)^{3/2}}{h^2}$	10^2–10^3
Vibrational (per normal mode)	1	$\dfrac{1}{1 - e^{-h\nu/\mathbf{k}T}}$	1–10
Restricted rotation	1	$\dfrac{(8\pi^3 I' \mathbf{k}T)^{1/2}}{h}$	1–10

m = mass of molecule
I = moment of inertia for linear molecule
$A, B,$ and C = moments of inertia for a nonlinear molecule, about three axes at right angles to one another
I' = moment of inertia for restricted rotation
ν = normal-mode vibrational frequency
$\mathbf{k}$ = Boltzmann constant
h = Planck's constant
T = temperature, °K

† For use in equilibrium expressions the rotational partition functions must be divided by symmetry numbers. In activated-complex theory, however, some difficulties arise if symmetry numbers are used in this way. The whole problem is discussed in some detail on pp. 65 to 75.

[A]^a[B]^b. . . . . The molecularity of a reaction is in fact equal to the number of reactant molecules that exist in the activated complex.

The rate constant of the reaction, the rate of which is given by Eq. (6), is given by

$$k = \frac{\mathbf{k}T}{h} \frac{Q_{\ddagger}}{Q_A Q_B} e^{-E_0/RT} \qquad (7)$$

The quantity $\mathbf{k}T/h$ which appears in these expressions is of great importance in rate theory; it has the dimensions of a frequency, and its value is about 6×10^{12} sec^{-1} at 300°K.

It is to be emphasized that when one states that the activated complexes are in equilibrium with the reactants one is referring only to complexes that are moving from left to right. This point is apt to lead to confusion, since it is customary, in calculating concentrations of activated complexes, to include those moving in both directions. Thus when one uses an equilibrium equation such as Eq. (2), where the Q's are the conventional partition functions, the concentration [X‡] includes complexes moving in both directions when the whole system is in equilibrium. In the absence of product molecules there are in fact only half of this number of activated complexes, all moving from left to right. The rate of reaction is, however, correctly given by multiplying [X‡] by the frequency ν, since a frequency is by definition equal to the frequency of crossing a barrier in one direction.

Derivation II In the second derivation, which is the one originally given by Eyring, the vibrational partition function corresponding to the coordinate of decomposition, instead of being retained in the form of $\mathbf{k}T/h\nu$, is replaced by a translational function. Figure 19 gives a schematic representation of the top of the potential-energy barrier, and it may be considered that all complexes lying within the length δ shown in the diagram are activated complexes; the actual value of δ will later be seen to be immaterial. The translational partition function corresponding to the motion of a particle of mass $m_{\ddagger}$ in a one-dimensional box of length δ is given by

$$q_t = \frac{(2\pi m_{\ddagger}\mathbf{k}T)^{1/2}}{h} \delta \qquad (8)$$

and if this expression is substituted for the vibrational partition function corresponding to the coordinate of decomposition, the resulting expression for the concentration of activated complexes is

$$[X^{\ddagger}] = [A][B] \frac{(2\pi m_{\ddagger}\mathbf{k}T)^{1/2}}{h} \delta \frac{Q_{\ddagger}}{Q_A Q_B} e^{-E_0/RT} \qquad (9)$$

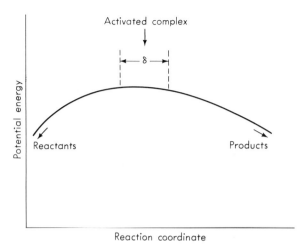

Fig. 19 The top of the potential-energy barrier. A flat portion at the top of the barrier, of length δ, is arbitrarily defined as comprising the activated state.

This again includes activated complexes moving in both directions and is therefore twice the number in which one is interested. The average velocity of the particles moving from left to right over the potential-energy barrier is given by kinetic theory as

$$\bar{\dot{x}} = \left(\frac{kT}{2\pi m_\ddagger}\right)^{1/2} \tag{10}$$

The frequency with which the complexes pass over the barrier from left to right is therefore given by this expression divided by δ, and the rate of reaction is the concentration of complexes multiplied by this frequency. The result is

$$v = [A][B] \frac{(2\pi m_\ddagger kT)^{1/2}}{h} \delta \left(\frac{kT}{2\pi m_\ddagger}\right)^{1/2} \frac{1}{\delta} \frac{Q_\ddagger}{Q_A Q_B} e^{-E_0/RT} \tag{11}$$

$$= [A][B] \frac{kT}{h} \frac{Q_\ddagger}{Q_A Q_B} e^{-E_0/RT} \tag{12}$$

Equation (12) is seen to be identical with Eq. (6).

The derivations given above appear to be different but they are basically the same. Derivation I regards the passage over the barrier as a very loose vibration, the other (II) as a free translation. If a particle is initially vibrating and the restoring force on it is gradually reduced to zero, the vibration ultimately becomes a translation. The partition function for a very loose vibration should therefore pass smoothly into that for a translation. Thus if the frequency corresponding to translational motion,

$$\nu = \left(\frac{\mathbf{k}T}{2\pi m_{\ddagger}}\right)^{1/2} \frac{1}{\delta} \tag{13}$$

is substituted in the expression $\mathbf{k}T/h\nu$, the result is $(2\pi m_{\ddagger}\mathbf{k}T)^{1/2} \delta/h$, which is the one-dimensional translational partition function for a particle in a length δ [Eq. (8)].

Derivation III   The third derivation, first given by Bishop and Laidler,[1] is a variant of derivation II, in which the equilibrium hypothesis is not explicitly made.

The curve RXP in Fig. 20 shows a cross section through a potential-energy surface, along the reaction coordinate and through the col, which corresponds to the activated state. Suppose that, hypothetically, the potential-energy surface did not descend toward the products after the activated state was reached, but after a short flat region rose as indicated by the dotted line XS. The state X then no longer corresponds to an activated complex, but is a state of an ordinary kind. An equilibrium

[1] D. M. Bishop and K. J. Laidler, *J. Chem. Phys.*, **42**:1688 (1965).

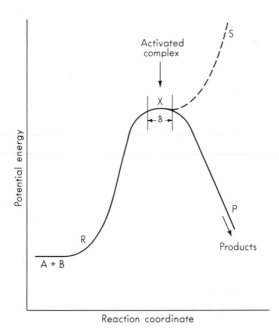

Fig. 20 A schematic potential-energy profile, relevant to derivation III of the rate equation.

concentration of these X states would be established. The concentration of the species X, defined as those species within the range δ, is given by

$$[X] = [A][B] \frac{Q^X}{Q_A Q_B} e^{-E_0/RT} \tag{14}$$

where Q^X is the complete partition function of the species X.

As in derivation II, we extract from the partition function Q^X the factor $kT/h\nu$, where ν is the frequency of motion in the region X; since this region is a flat one, this frequency is actually ·equal to the average kinetic-theory velocity in one dimension $(kT/2\pi m_{\ddagger})^{1/2}$ divided by the distance δ:

$$\nu = \left(\frac{kT}{2\pi m_{\ddagger}} \right)^{1/2} \frac{1}{\delta} \tag{15}$$

Equation (14) can be written as

$$[X] = [A][B] \frac{kT}{h\nu} \frac{Q_X}{Q_A Q_B} e^{-E_0/RT} \tag{16}$$

where Q_X is the partition function Q^X from which the factor kT/h has been removed.

Since the state X is in equilibrium with $A + B$, the number of species per second passing through the state from left to right must be equal to the number per second passing from right to left. The number passing from right to left per second, v_r, is given by

$$v_r = \nu[X] \tag{17}$$

The number passing from left to right, i.e., into the state X from $A + B$, is equal to this and by use of (16) is given by

$$v_1 = [A][B] \frac{kT}{h} \frac{Q_X}{Q_A Q_B} e^{-E_0/RT} \tag{18}$$

So far we have been considering systems moving on the surface RXS. Suppose that we now consider the true surface, which is identical from R to X, but differs in descending beyond X. The rate with which systems enter the state X, which is now the activated state, is obviously the same as that derived above [Eq. (18)] since the species do not "know," when arriving at X, whether the surface is then going to rise or fall. If it is now postulated that every species that has reached the activated state has reached the point of no return, Eq. (18) gives the rate of reaction.

This last derivation was not based upon any special assumption of equilibrium between initial and activated states. The device of considering first the surface RXS was used to arrive at an expression for the rate with which the complexes were formed. The only sense in which an equilibrium assumption is involved is that it is tacitly assumed that the equilibrium distribution between the various energy levels in the reactants is not significantly disturbed by the occurrence of reaction. More elaborate treatments, referred to later (pages 211 and 217), have shown that this assumption is valid for reactions occurring at ordinary speeds but will be in error for very fast reactions.

It is easy to show on the basis of the expressions derived above that there does exist an equilibrium concentration of activated complexes. The rate of entrance of reactant species into the activated state is given by Eq. (18), and this is equal to the rate of exit, which is $\nu[X^{\ddagger}]$. The central point is that ν [Eq. (15)] is independent of the subsequent form of the potential-energy surface (Fig. 20). When these rates are equated,

$$\nu[X^{\ddagger}] = [A][B]\frac{\mathbf{k}T}{h}\frac{Q_{\ddagger}}{Q_A Q_B} e^{-E_0/RT} \tag{19}$$

whence

$$[X^{\ddagger}] = [A][B]\frac{\mathbf{k}T}{h\nu}\frac{Q_{\ddagger}}{Q_A Q_B} e^{-E_0/RT} \tag{20}$$

This is the equilibrium expression [cf. Eq. (5)].

TEMPERATURE DEPENDENCE OF THE FREQUENCY FACTOR

The three derivations given above have led to the rate equation

$$k = \frac{\mathbf{k}T}{h}\frac{Q_{\ddagger}}{Q_A Q_B} e^{-E_0/RT} \tag{21}$$

The problem of evaluating the rate constant has thus been reduced to that of evaluating the partition functions for the initial and activated states. The main difficulty is in arriving at a decision about the nature of the activated complexes. In principle, the structures of activated complexes can be determined by the methods of quantum mechanics, as discussed in the last chapter. Even in the simplest cases the procedure is laborious and not very accurate. In many cases, however, the structure of the activated complex can be estimated with sufficient accuracy from our general knowledge of molecular structure.

It follows from the above formulations that the frequency factor A [see Eq. (1), Chap. 1] is not necessarily independent of temperature. Thus in Eq. (21) the temperature appears in $\mathbf{k}T/h$ and also in the partition functions. Except for the vibrational factors, which are close to unity in most cases and do not depend very much on the temperature, the partition functions involve the temperature to a simple power (see Table

3), so that the rate constant for a reaction can be expressed in the form

$$k = aT^n e^{-E_0/RT} \tag{22}$$

The value of the exponent n depends upon the form taken by the partition functions. The fact that plots of the logarithm of k against the reciprocal of the temperature are usually linear for elementary reactions is due, as discussed in Chap. 1, to the much stronger temperature dependence of the exponential part than of the preexponential part. When the rate equation is of the above form, the apparent energy of activation is temperature-dependent; the way in which it varies with temperature may be deduced as follows.

The experimental energy of activation E_{exp} is defined by

$$\frac{d \ln k}{dT} = \frac{E_{exp}}{RT^2} \tag{23}$$

since the experimental activation energy is determined by plotting the logarithm of k against the reciprocal of the absolute temperature. Differentiation of the logarithmic form of Eq. (22) gives

$$\frac{d \ln k}{dT} = \frac{n}{T} + \frac{E_0}{RT^2} = \frac{E_0 + nRT}{RT^2} \tag{24}$$

Comparison of these two equations leads to the relationship

$$E_{exp} = E_0 + nRT \tag{25}$$

At the absolute zero the energy of activation is the difference between the zero-point levels in the initial and the activated states, whereas at any other temperature it is the difference between the average energies of the reactants and of the activated complex. Unless n is very large or E_0 is very small, it may be difficult experimentally to detect this temperature dependence of the energy of activation. Except for some trimolecular reactions, considered in Chap. 5, this has rarely been done for gas reactions, but there are some reactions in solution where the temperature variations are very large and have been detected.

THERMODYNAMICAL FORMULATION OF REACTION RATES

It is sometimes convenient to express the rate constants of reactions in terms of thermodynamical functions rather than partition functions; this is often done with reactions in solution, since partition functions of species in the liquid phase are very difficult to evaluate. The thermodynamical formation of rate constants is based on the fact that the equilibrium between reactants and activated complexes may be expressed in terms of thermodynamical functions as well as by using partition functions.

The equilibrium constant for the process $A + B \rightleftharpoons X$ may be written as

$$K^{\ddagger} = \left(\frac{[X^{\ddagger}]}{[A][B]}\right)_{eq} = \frac{Q_{\ddagger}}{Q_A Q_B} e^{-E_0/RT} \tag{26}$$

Strictly speaking the partition function of the activated complex that appears in this expression should be $Q^{\ddagger}$ instead of $Q_{\ddagger}$. The equilibrium constant $K^{\ddagger}$ is therefore a special type of equilibrium constant. Comparison of Eqs. (26) and (7) leads to

$$k = \frac{kT}{h} K^{\ddagger} \tag{27}$$

and if $K^{\ddagger}$ is expressed in terms of $\Delta G^{\ddagger}$, the increase in standard Gibbs free energy in the passage from the initial state to the activated state, the result is

$$k = \frac{kT}{h} e^{-\Delta G^{\ddagger}/RT} \tag{28}$$

This free energy of activation, $\Delta G^{\ddagger}$, may be expressed in terms of an entropy and a heat of activation, that is, as $\Delta H^{\ddagger} - T \Delta S^{\ddagger}$:

$$k = \frac{kT}{h} e^{\Delta S^{\ddagger}/R} e^{-\Delta H^{\ddagger}/RT} \tag{29}$$

This equation was first derived in 1936 by Wynne-Jones and Eyring.[1] If k is expressed in liters mole^{-1} sec^{-1}, the standard state for the free energy and the entropy of activation is 1 mole/liter.

Equation (29) may be expressed in a form that involves the experimental energy of activation E_{exp} instead of the heat of activation $\Delta H^{\ddagger}$. Since $K^{\ddagger}$ is a concentration equilibrium constant, its variation with temperature is given by the equation

$$\frac{d \ln K^{\ddagger}}{dT} = \frac{\Delta E^{\ddagger}}{RT^2} \tag{30}$$

where $\Delta E^{\ddagger}$ is the increase in energy in passing from the initial state to the activated state. Differentiation of the logarithmic form of Eq. (27) gives

$$\frac{d \ln k}{dT} = \frac{1}{T} + \frac{d \ln K^{\ddagger}}{dT} \tag{31}$$

and together with Eq. (30) this gives

$$\frac{d \ln k}{dT} = \frac{1}{T} + \frac{\Delta E^{\ddagger}}{RT^2} = \frac{RT + \Delta E^{\ddagger}}{RT^2} \tag{32}$$

[1] W. F. K. Wynne-Jones and H. Eyring, *J. Chem. Phys.*, **3**:492 (1935).

Comparison of this equation with Eq. (23) leads to

$$E_{exp} = RT + \Delta E^{\ddagger} \tag{33}$$

The relationship between $\Delta E^{\ddagger}$ and $\Delta H^{\ddagger}$ is

$$\Delta H^{\ddagger} = \Delta E^{\ddagger} + P \Delta V^{\ddagger} \tag{34}$$

where $\Delta V^{\ddagger}$ is the increase in volume in going from the initial state to the activated state. Substitution of this in Eq. (33) gives

$$E_{exp} = \Delta H^{\ddagger} - P \Delta V^{\ddagger} + RT \tag{35}$$

For unimolecular reactions there is no change in the number of molecules as the activated molecule is formed, and $\Delta V^{\ddagger}$ is therefore zero; $\Delta V^{\ddagger}$ is also small for reactions in solution. In these cases

$$E_{exp} = \Delta H^{\ddagger} + RT \tag{36}$$

and the rate equation may therefore be written as

$$k = \frac{kT}{h} e^{\Delta S^{\ddagger}/R} e^{-(E_{exp}-RT)/RT} \tag{37}$$

or as

$$k = e \frac{kT}{h} e^{\Delta S^{\ddagger}/R} e^{-E_{exp}/RT} \tag{38}$$

For gas reactions the general relationship is

$$P \Delta V^{\ddagger} = \Delta n^{\ddagger} RT \tag{39}$$

where $\Delta n^{\ddagger}$ is the increase in the number of molecules when the activated complex is formed from the reactants. In a bimolecular reaction, for example, two molecules become one, so that $\Delta n^{\ddagger}$ is equal to -1; in this case the experimental energy of activation is related to the heat of activation by the relationship

$$E_{exp} = \Delta H^{\ddagger} + 2RT \tag{40}$$

From this it follows that the rate constant may be written as

$$k = e^2 \frac{kT}{h} e^{\Delta S^{\ddagger}/R} e^{-E_{exp}/RT} \tag{41}$$

For agreement with the simple hard-sphere theory of collisions the entropy of activation, relative to the units of liters mole^{-1} sec^{-1}, must be about -12 cal deg^{-1} mole^{-1}. If, on the other hand, the standard state is 1 mole/cc, that is, if the rate-constant units are cubic centimeters mole^{-1} sec^{-1}, the entropy of activation must be approximately zero for agreement with the collision theory. For this reason the units of cubic centimeters mole^{-1} sec^{-1} are perhaps more convenient.

4

Some Special Topics

The present chapter deals with some problems that arise in applying the activated-complex rate equation to actual reactions.

TRANSMISSION COEFFICIENT

It is sometimes necessary to correct the rate equation by a factor that takes account of the possibility that not every activated complex becomes transformed into the particular reaction product with which one is concerned. This factor is referred to as the *transmission coefficient* and is usually denoted by the symbol κ.

The dynamical calculations that have recently been made on a number of reactions, including $H + H_2$, have indicated a *direct* or *impulsive* type of mechanism. That is to say, the systems that reach the activated state pass directly through it, and only on very rare occasions do they return to the initial state. In such reactions the activated state is therefore a point of no return. When this is the case, the transmission coefficient is very close to unity.

There is some evidence, to be considered in Chap. 7, that a few reactions occur by an *indirect* (sometimes called a *complex*) mechanism, as noted in Chap. 3.   In these there is formed a complex which undergoes several vibrations and thus has a reasonable chance of returning to the initial state.   The transmission coefficient is then less than unity.   It appears, although this is still not entirely certain, that this only happens in reactions for which the potential-energy surface has a basin, corresponding to a more or less stable complex which can undergo a number of vibrations.

Another situation in which the transmission coefficient may be less than unity is in bimolecular atom-combination reactions in the gas phase:

$$A + A \rightarrow A_2$$

When two atoms collide in the gas phase an activated complex may be said to be formed on every collision.   However, the resulting molecule still contains, largely as vibrational energy, the energy of the initial atoms and therefore decomposes in the period of its first vibration. There is a small possibility that the species will lose energy by radiation, but apart from this the transmission coefficient is zero.   If, on the other hand, the complex is formed in the presence of a third body, the energy may be removed and the product stabilized; the transmission coefficient may then approach unity.   Low transmission coefficients may also be found in second-order radical-combination reactions.   Here, however, energy may pass into other normal modes of vibration.   The transmission coefficient in fact approaches unity as the radicals become more and more complex.

The transmission coefficient may also be appreciably less than unity in those reactions where there is a crossing of potential-energy surfaces, so that there are alternative reaction paths.   In some cases, for example, the activated state corresponds to a crossing point, as represented schematically in Fig. 21.   The extent of resonance splitting at the crossing of two surfaces depends in a marked manner on whether the electronic state corresponding to the two surfaces are of the *same species* (Fig. 21a) or of *different species* (Fig. 21b).   States of the same species are those with the same symmetry properties and the same Λ and S values; the resonance splitting is large and the probability of crossing from the upper to the lower surface is very small.   With states of different species the probability of transition is close to unity.   As a general principle it can be said that systems resist any change in spin angular momentum and hence in multiplicity.   This principle is often referred to as the *Wigner spin-conservation rule;*[1] some of its kinetic implications have been discussed elsewhere.[2]

[1] E. Wigner, *Nachr. Ges. Wiss. Göttingen* **1927**:327.

[2] K. J. Laidler, "The Chemical Kinetics of Excited States," Clarendon Press, Oxford, 1955.

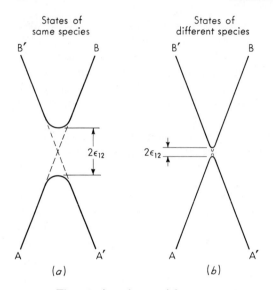

States of
same species

States of
different species

Fig. 21 The crossing of potential-energy surfaces.

For states of the same species no quantitative treatment of transition probabilities appears to have been given. The probability that the system remains on the lower surface (i.e., passes from A to A′) is very close to unity. For states of different species the probability of crossing (i.e., passing from A to B) was independently given by Landau,[1] Zener,[2] and Stueckelberg[3] as

$$P = \exp \frac{-4\pi^2 \varepsilon_{12}^{\,2} \mathbf{k} T}{hv|s_1 - s_2|} \tag{1}$$

Here ε_{12} is the resonance integral (equal to one-half of the separation when the curves are closest together), v is the velocity with which the system passes the point of closest approach of the surfaces, and $|s_1 - s_2|$ is the absolute magnitude of the difference of the slopes with which the surfaces would cross if resonance were not taken into account. This equation is intended to apply only if ε_{12} is very small compared with the potential energy of the system when it reaches the crossing point. It is seen from Eq. (1) that the probability of crossing approaches unity as ε_{12} approaches zero and as the velocity v approaches infinity. If the velocity is very small, the probability of crossing is large. The transmission coefficient for a process A → A′ therefore tends to be low, perhaps of the order of 10^{-4}, for states of different species.

[1] L. Landau, *Phys. Z. Sowjetunion*, **1**:88 (1932); **2**:46 (1932).
[2] C. Zener, *Proc. Roy. Soc.*, *(London)* **A137**:696 (1932); **A140**:666 (1933).
[3] E. G. C. Stueckelberg, *Helv. Phys. Acta*, **5**:369 (1932).

However, the Landau-Zener-Stueckelberg formula is based on a number of rather unrealistic assumptions, and it cannot be expected to give more than a qualitative idea of transition probabilities. Some of these assumptions have been discussed and attempts have been made to remove them by Bates[1] and by Coulson and Zalewski.[2] The problem is one of considerable difficulty, and these later treatments are rather formal; they have still not been applied to any actual systems. There is considerable need for further theoretical work in this important field.

QUANTUM-MECHANICAL TUNNELING

According to classical mechanics the occurrence of a chemical reaction must correspond to the passage of the system over the top of a potential-energy barrier. Quantum-mechanical theory, on the other hand, admits the possibility that a system having less energy than is required to surmount the barrier may nevertheless pass from the initial to the final state; it is said to *tunnel* or *leak* through the potential-energy barrier. Tunneling is most important for a particle of small mass, and when the barrier is low and narrow. Quantum-mechanical tunneling plays a particularly important role in electron-transfer reactions. It is much less important, but significant, for reactions involving the transfer of protons and deuterons, either as ions or with their accompanying electrons. For heavier atoms or ions, tunneling is of negligible importance. Tunneling is considerably more important with protons than with deuterons, and this must clearly be taken into account when one is dealing with hydrogen-deuterium isotope effects.

The quantitative treatment of tunneling, with its application to the experimental results, is a matter of some difficulty, and the situation is still by no means clear. It is relatively easy to obtain explicit expressions for the probability of tunneling through one-dimensional barriers of artificial shapes such as rectangles[3] and triangles[4] (Fig. 22). Wigner[5] considered a one-dimensional parabolic barrier, shown in the figure. If this barrier were turned upside down, the system would vibrate with a certain frequency ν; in fact, with the barrier as it is, the vibrational frequency is an imaginary quantity. If ν is small, the barrier is a broad one and the probability of tunneling is low; for this case Wigner has given the approxi-

[1] D. R. Bates, *Proc. Roy. Soc. (London)*, **A257**:22 (1960).

[2] C. A. Coulson and K. Zalewski, *Proc. Roy. Soc. (London)*, **A268**:437 (1962).

[3] N. F. Mott and I. N. Sneddon, "Wave Mechanics and its Applications," pp. 15–17, Clarendon Press, Oxford, 1948.

[4] G. Gamow, *Z. Physik*, **51**:204 (1928).

[5] E. P. Wigner, *Z. Physik, Chem. (Leipzig)*, **B19**:903 (1932).

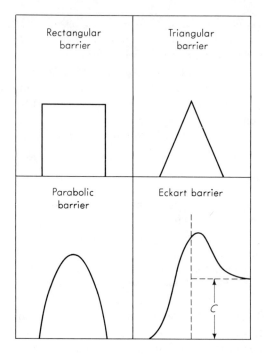

Fig. 22 Artificial energy barriers of various kinds.

mate expression[1]

$$Q = 1 + \frac{1}{24}\left(\frac{h|\nu|}{\mathbf{k}T}\right)^2 \tag{2}$$

for the tunneling correction; this is the factor that must be multiplied into the rate equation to allow for tunneling. A set of approximate expressions covering a wider range of parabolic barriers has been given by Bell.[2]

A one-dimensional barrier shape that corresponds more closely to physical reality and for which a solution of the quantum-mechanical problems is possible was proposed by Eckart.[3] Its form is shown in Fig. 22; the variation of potential, V, with reaction coordinate x is given by

$$V = \frac{C\eta}{1+\eta} + \frac{D\eta}{(1+\eta)^2} \tag{3}$$

where

$$\eta = e^{2\pi x/b} \tag{4}$$

[1] $|\nu|$ represents the modulus of the imaginary frequency ν, so that $|\nu|^2$ is positive and $Q > 1$.

[2] R. P. Bell, *Trans. Faraday Soc.*, **55**:1 (1959); also *Proc. Roy. Soc. (London)*, **A139**:466 (1933); **A148**:241 (1935); **A158**:128 (1937).

[3] C. Eckart, *Phys. Rev.*, **35**:1303 (1930).

and C, D, and b are constants. The barrier may be unsymmetrical, as shown, or symmetrical. The expressions for the permeability of such barriers and for the correction factor Q are rather complicated in form and involve integrals which have to be evaluated numerically. Shavitt[1] has published, in tabular form, values of Q for a variety of conditions for the parabolic and Eckart barriers, and similar tables have been given by Johnston and Rapp.[2] Approximate solutions, in closed form, for the unsymmetrical Eckart barrier have also been given by Shin.[3]

Practically all of the treatments that have been given so far are oversimplified in that they relate to a two-dimensional tunneling process. A system crossing a potential-energy barrier can in fact tunnel from various positions which are away from the minimum-reaction path; the treatment of this situation is, however, very difficult. Mortensen and Pitzer[4] have provided evidence that the simple one-dimensional treatments tend to *underestimate* the extent of tunneling.

It is not an entirely straightforward matter to obtain good experimental evidence for quantum-mechanical tunneling. Theory predicts that when tunneling occurs certain effects will be observed; however, these effects may sometimes be due to other causes, so that the evidence may be ambiguous. The main effects predicted on the basis of tunneling are:

1. There will be a deviation from linearity of the kind shown schematically in Fig. 23. The reason for this is that according to all of the theoretical treatments tunneling becomes relatively more important as the temperature is lowered. Equation (2) indeed predicts that Q will be larger, the lower the temperature; since the rate of reaction *over* the barrier normally increases strongly with the temperature, the relative importance of tunneling is much greater at lower temperatures.

Behavior of the kind shown in Fig. 23 has been observed in a number of cases. However, this type of behavior will also be observed if two reactions are occurring simultaneously (the one of higher activation energy becoming relatively more important at the higher temperatures); care must therefore be taken to eliminate this possibility. Evidence of the occurrence of tunneling, on the basis of nonlinear Arrhenius plots, has been put forward for the reactions

$$H + H_2 \rightarrow H_2 + H$$

and

$$D + H_2 \rightarrow DH + H$$

[1] I. Shavitt, *J. Chem. Phys.*, **31**:1359 (1959).

[2] H. S. Johnston and D. Rapp, *J. Am. Chem. Soc.*, **83**:1 (1961).

[3] H. Shin, *J. Chem. Phys.*, **39**:2934 (1963).

[4] E. M. Mortensen and K. S. Pitzer, *Chem. Soc. (London) Spec. Pub.* **16**:57 (1962).

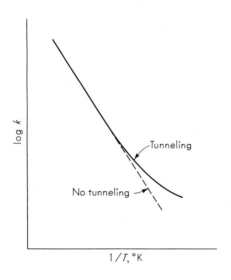

Fig. 23 Schematic Arrhenius plot for a reaction in which quantum-mechanical tunneling occurs.

by LeRoy and coworkers.[1]   This work is discussed in more detail in the next chapter.   A number of reactions in solution have also been found to show nonlinear Arrhenius behavior of the type shown in Fig. 23.   The most thoroughly investigated example is the base-catalyzed bromination, in heavy water, of the cyclic ketone 2-carbethoxycyclopentanone; this reaction is a base-catalyzed reaction and is undoubtedly controlled by proton transfer from the substrate to the base.   For this reaction catalyzed by fluoride ions and other bases Bell and coworkers[2] found a very marked deviation from linearity in the Arrhenius plot, an effect that cannot be explained in other ways.   At 20°C, for example, the rate was 75 percent faster than that calculated by extrapolating the linear Arrhenius plot obtained at higher temperatures.   The experiments were interpreted in terms of the equations for the penetration of a parabolic barrier; the width of this at its base was calculated as 1.17 Å, and the height is about 20 percent greater than the observed activation energy.   Similar deviation from the Arrhenius behavior has been found by Caldin and coworkers[3] for the reactions of the trinitrobenzyl ion with acetic acid and with

[1] W. R. Schulz and D. J. LeRoy, *J. Chem. Phys.*, **42**:3869 (1965); B. A. Ridley, W. R. Schulz, and D. J. LeRoy, *J. Chem. Phys.* **44**:3344 (1966); W. R. Schulz and D. J. LeRoy, *Can. J. Chem.*, **42**:2480 (1964).

[2] R. P. Bell, J. A. Fendley, and J. R. Hulett, *Proc. Roy. Soc. (London)*, **A235**:453 (1956); J. R. Hulett, *Proc. Roy. Soc. (London)*, **A251**:274 (1959).

[3] E. F. Caldin and E. Harbron, *J. Chem. Soc.*, **1962**: 3454; E. F. Caldin, M. Kasparian, and G. Tomalin, *Trans. Faraday Soc.*, **64**:2802 (1968); E. F. Caldin and G. Tomalin, *Trans. Faraday Soc.*, **64**:2814, 2823 (1968); E. F. Caldin, "Fast Reactions in Solution," p. 273, Blackwell, Oxford, 1964.

hydrofluoric acid, and for the reaction of 4-nitrobenzyl cyanide with ethoxide ions. The work on these three reactions was done down to -115, -90, and $-124°C$; the corresponding deviations of the rate constants at the lowest temperatures, from the values expected from extrapolation of the Arrhenius plots, were 45, 120, and 100 percent. These deviations are far outside the experimental error, and no explanation other than tunneling is satisfactory.

2. The second effect predicted to occur when there is tunneling is related to the first; this is that frequency factors will be abnormally small. It can be seen from Fig. 23 that, since log A is log k extrapolated to a zero value of $1/T$, the frequency factor at low temperatures, where tunneling is important, will be low. The result is that, for example, the frequency factor of a proton transfer may be considerably less than that for the corresponding deuteron transfer; this would be difficult to explain on other grounds. In the work on the bromination of 2-carbethoxycyclo-pentanone, A_D/A_H was found to be 24 at low temperatures; this is strong evidence for tunneling, especially taken in conjunction with the nonlinear Arrhenius behavior.

3. Apart from these frequency-factor differences, abnormally large isotope effects are predicted where tunneling occurs, the lighter isotope reacting relatively more rapidly than can be explained without taking tunneling into account. This kind of evidence is never very clear-cut, since the treatment of isotope effects presents some difficulty and depends strongly on the form of the potential-energy surface; as seen in Chap. 2, such surfaces are never known with certainty. In the following chapter the isotopic effects observed in several gas reactions are discussed, and it is shown that there is some evidence for tunneling.

THE REACTION COORDINATE

In the derivation of the rate equation (page 46) it was noted that one of the motions of the activated complex is of a special character in that it corresponds to the free passage of the system over the barrier. The nature of this motion depends upon the form of the potential-energy surface in the region of the activated complex. Two cases of particular interest are represented schematically in Fig. 24, which relates to a linear triatomic complex A—B—C; for such a complex there are four modes of vibration, of which two are for stretching and two (not represented in the figure) are for bending. In Fig. 24a the valleys are arranged symmetrically, so that the passage of the system across the col corresponds to the antisymmetrical vibration of the complex of frequency ν_2. The correct procedure in treating a reaction of this type is to omit the vibration of frequency ν_2 from the partition function and to consider only the vibra-

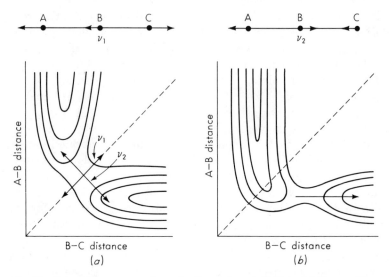

Fig. 24 Potential-energy surfaces for the system A-B-C; in (a) the reaction coordinate corresponds to the antisymmetric vibration of frequency ν_2; in (b) it corresponds to a simple stretching of the B—C bond.

tion of frequency ν_1 and the two bending vibrations. In such a case the reaction coordinate in the activated state may be said to correspond to the unsymmetric vibration. For a more complicated activated complex the passage across the barrier may in some instances correspond to one particular vibration, and the corresponding partition function would then be omitted in the rate expression.

The potential-energy surfaces may, however, be of such a form that the reaction coordinate does not correspond to one normal mode of vibration. In Fig. 24b, for example, the surfaces have been drawn in such a way that the free passage over the col corresponds to an extension of the B—C bond, the A—B distance remaining constant; the reaction coordinate is therefore the B—C distance. The procedure in this case is to replace the ordinary normal-mode analysis of this problem by one in which the B—C distance is held fixed. For the triatomic case there would therefore be the ordinary bending motions and also the motion of the atom A along the line of centers in B—C. The frequency of this stretching vibration is

$$\nu^* = \frac{1}{2\pi} \sqrt{\frac{k_{AB}}{m^*}} \tag{5}$$

where k_{AB} is the force constant related to the curvature of the potential-energy surface parallel to the A—B axis, and m^* is the reduced mass, given by

$$\frac{1}{m^*} = \frac{1}{m_A} + \frac{1}{m_B + m_C} \tag{6}$$

For activated complexes containing more than three atoms the procedure is similar, but more difficult to carry out. If the reaction coordinate can be identified with a normal mode, the ordinary normal-mode analysis reveals all the frequencies, and the appropriate one is omitted. If the reaction coordinate corresponds to a single interatomic distance, the procedure is to carry out a normal-mode analysis with this distance held constant. If it corresponds to a linear combination of two or more normal modes, a normal-mode analysis is carried out with this linear combination held fixed.

The precise choice of reaction coordinate is not of great importance if one is content with an approximate estimate of a rate constant; vibrational frequencies of molecules do not vary widely, particularly for stretching vibrations, which are usually the important ones in connection with the reaction coordinate,[1] and the rate constant does not depend strongly upon which frequency is omitted. For exact calculations, however, the choice of reaction coordinate is of great importance and can only be made on the basis of a potential-energy surface. These cannot be constructed in a reliable manner even for the simplest of reactions, so that there is clearly a serious difficulty in making exact calculations. This situation leads to a particular difficulty, and one that has not always been appreciated, in dealing with kinetic isotope effects.

STATISTICAL FACTORS

In reactions involving molecules having symmetry some complications arise. These have led to considerable confusion in the past, and the problem will be dealt with in some detail in the present section.

Equilibrium constants   In the evaluation of equilibrium constants from partition functions it is customary to divide the vibrational partition function for each of the molecules involved by its *symmetry number* σ. The symmetry number is found by imagining all identical atoms to be labeled and counting the number of different but equivalent arrangements that can be made by rotating (but not reflecting) the molecule. Thus if the atoms in the hydrogen molecule are labeled 1 and 2, the following two arrangements are possible:

$$H^1\text{—}H^2 \qquad H^2\text{—}H^1$$

---

[1] The most common exception is found in cis-trans isomerization, where the reaction coordinate may be simply the angle of twist.

The symmetry number is therefore 2, as it is in the water molecule:

$$
\text{H}^1 \diagup \overset{\text{O}}{} \diagdown \text{H}^2 \qquad \text{H}^2 \diagup \overset{\text{O}}{} \diagdown \text{H}^1
$$

The symmetry number of ammonia (nonplanar) is 3, while that of the planar methyl radical is 6:

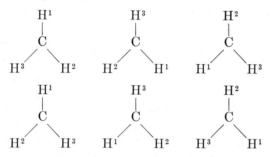

It can easily be verified that the symmetry number of methane is 12, as is that of benzene.

The reason for the necessity of taking symmetry into account in calculating equilibrium constants is seen with the two reactions

1. $H_2 + Cl \rightleftharpoons HCl + H$
2. $HD + Cl \rightleftharpoons HCl + D$

If the partition functions *without symmetry numbers* are denoted by Q^0, the equilibrium constants for these two reactions would be

$$
K_1 = \frac{2Q^0_{HCl}Q_H{}^0}{Q_{H_2}{}^0 Q_{Cl}} e^{-E_0/RT} \tag{7}
$$

and

$$
K_2 = \frac{Q^0_{HCl}Q_D{}^0}{Q_{HD}{}^0 Q_{Cl}{}^0} e^{-E_0/RT} \tag{8}
$$

The 2 appears in the numerator of (7) because the partition function for H_2, Q_{H_2} is $Q_{H_2}{}^0$ divided by the symmetry number 2. It follows that if the Q^0's were unaffected by the replacement of H by D (for example, if $Q_{HD}{}^0 = Q_{H_2}{}^0$) and the E_0's are the same, K_1 would be twice K_2. That is to say, there is a statistical factor of 2 favoring reaction 1.

The origin of this statistical factor is simply that, if we label the H atoms in H_2, we can envisage reaction 1 as producing HCl in two different ways,

$$
\text{H}^1 - \text{H}^2 + \text{Cl} \rightarrow \text{H}^1 - \text{Cl} + \text{H}^2
$$

and

$$
\text{H}^1 - \text{H}^2 + \text{Cl} \rightarrow \text{H}^2 - \text{Cl} + \text{H}^1
$$

There is therefore a double chance of getting HCl, which is not so if we specify in reaction 2 that it is HCl we are producing and not DCl. If reaction 2 were expanded to include the formation of HCl and DCl indiscriminately,

$$HD + Cl \underset{\searrow}{\overset{\nearrow}{}} \begin{array}{l} HCl + D \\[1em] DCl + H \end{array}$$

the overall equilibrium constant with HCl and DCl treated as identical would be the same as for reaction 1. If, however, we specify that our equilibrium constant refers to the formation of HCl only, the value for reaction 2 will be half that for reaction 1 (again, we are now concerned only with the statistical factor and are treating the partition functions as otherwise unaffected by isotopic substitution).

It follows from this argument that instead of using symmetry numbers in calculating equilibrium constants from partition functions we could multiply by the ratio of appropriate *statistical factors*. These are counted by labeling identical atoms and seeing how many equivalent product species are obtained.[1] This has already been seen to work for reaction 1. A more complicated example is

$$H + CH_4 \rightleftharpoons H_2 + CH_3 \text{ (planar)}$$

The use of symmetry numbers leads to

$$K = \frac{(Q_{H_2}{}^0/2)(Q_{CH_3}^0/6)}{Q_H{}^0 Q_{CH_4}^0/12} e^{-E_0/RT} \tag{9}$$

$$= \frac{Q_{H_2}{}^0 Q_{CH_3}^0}{Q_H{}^0 Q_{CH_4}^0} e^{-E_0/RT} \tag{10}$$

If we label the hydrogen atoms on the left-hand side,

$$H^1 + \quad \begin{array}{c} H^2 \\ | \\ C \\ \nearrow \; | \; \searrow \\ H^3 \; H^5 \; H^4 \end{array}$$

---

[1] If two identical molecules are involved in a reaction, the statistical factor must be taken as the number of equivalent products formed divided by 2; thus, for the reaction

$$\underset{\sigma=1}{HI} + \underset{\sigma=1}{HI} \underset{r=2}{\overset{l=1/2}{\rightleftharpoons}} \underset{\sigma=2}{H_2} + \underset{\sigma=2}{I_2}$$

the statistical factors are as indicated, and their ratio is the ratio of symmetry numbers (1:4).

the following four sets of products can be obtained:

$$H^1H^3 + CH^3H^4H^5$$
$$H^1H^3 + CH^2H^4H^5$$
$$H^1H^4 + CH^2H^3H^5$$
$$H^1H^5 + CH^2H^3H^4$$

The statistical factor l for the reaction from left to right is thus 4. Similarly, we can label the reaction from right to left as follows:

$$H^1H^2 + CH^3H^4H^5$$

and find that four forms of $H + CH_4$ are produced:

The statistical factor r for this back reaction is thus 4. The correct equilibrium constant is thus obtained for this reaction if we use the Q^0 values and multiply by l/r, which is unity in the present case.

It can be proved[1] that *for any reaction*, such as

$$A + B \overset{l}{\underset{r}{\rightleftharpoons}} C + D$$

the ratio l/r of statistical factors is equal to the ratio $\sigma_A\sigma_B/\sigma_C\sigma_D$ of symmetry numbers. The use of the ratio l/r in calculating an equilibrium constant therefore gives the same answer as using the symmetry numbers.

[1] D. M. Bishop and K. J. Laidler, *J. Chem. Phys.*, **42**:1688 (1965).

The proof of the theorem is as follows.   Consider first the reaction

$$A + B \underset{r'}{\overset{l'}{\rightleftharpoons}} X$$

where X is a hypothetical intermediate species having no rotational symmetry elements ($\sigma_X = 1$).   The formation of X from A and B can be thought of as changing the positions in space of certain atoms of A and B to form X; since X has no rotational symmetry there are r' ways of doing this.   We can, however, rotate A and B and obtain $\sigma_A \sigma_B$ equivalent arrangements, the difference between them being that the atoms in A and B are labeled differently.   Since $\sigma_X = 1$, each arrangement will lead to a different product X; thus the number l' of different products formed is[1]

$$l' = r'\sigma_A\sigma_B \tag{11}$$

Any reaction can be considered to occur through some hypothetical intermediate X which has no rotational symmetry elements,

$$A + B \underset{r'}{\overset{l'}{\rightleftharpoons}} X \underset{r''}{\overset{l''}{\rightleftharpoons}} C + D$$

It follows that

$$\frac{l'}{r'} = \sigma_A\sigma_B \tag{12}$$

and

$$\frac{l''}{r''} = \frac{1}{\sigma_C\sigma_D} \tag{13}$$

The overall statistical factor l is equal to $l'l''$, and $r = r'r''$; it follows that

$$\frac{l}{r} = \frac{l'l''}{r'r''} = \frac{\sigma_A\sigma_B}{\sigma_C\sigma_D} \tag{14}$$

There is much to be said, in calculating equilibrium constants from partition functions, for using statistical factors instead of symmetry numbers.   It is if anything easier to count statistical factors than symmetry numbers.   More important, statistical factors give one an intuitive grasp of the reason why these corrections have to be made; symmetry numbers are not easily understood in this way, and many people use them without knowing why.   Finally, there is one situation where the use of symmetry numbers can easily lead to the wrong answer unless care is taken.   This is when a reaction introduces or removes an asymmetric center.   The

---

[1] If A and B are identical, there will be $2\sigma_A\sigma_B$ equivalent arrangements, but by definition l in this case is equal to one-half the number of different X molecules; l therefore remains equal to $\sigma_A\sigma_B$.

simplest example is the addition of a hydrogen atom to an unsymmetrically substituted methyl radical, which has a planar configuration,

Since two enantiomers are formed, $l = 2$ but $r = 1$; the correct equilibrium constant is thus

$$K = 2 \frac{Q^0_{\mathrm{CPQRH}}}{Q_\mathrm{H}{}^0 Q^0_{\mathrm{CPQR}}} e^{-E_0/RT} \tag{15}$$

However, all the species involved have symmetry numbers of unity, so that a straightforward application of the symmetry-number procedure would not lead to the 2 in this expression. The reason is, of course, that the symmetry number method is concerned with the equilibrium constant involving *one* enantiomer only. Gold[1] has suggested, in order to overcome this difficulty, that when a molecule is a mixture of enantiomers, its symmetry number should formally be taken as $\frac{1}{2}$. This certainly gives the right answer; however, it is easy, in evaluating symmetry numbers, to overlook the existence of enantiomers. The statistical-number method completely avoids this difficulty.

Rate equations　Some subtleties arise in the application of these ideas to rate equations, and confusion has existed for some time. The present discussion is based on treatments given by Bishop and Laidler[2] and by Murrell and Laidler.[3]

It was originally assumed that correct values for rate constants would be obtained by simply introducing the appropriate symmetry numbers into the partition functions:

$$k = \frac{\mathbf{k}T}{h} \frac{Q_\ddagger{}^0/\sigma_\ddagger}{(Q_\mathrm{A}{}^0/\sigma_\mathrm{A})(Q_\mathrm{B}{}^0/\sigma_\mathrm{B})} e^{-E_0/RT} \tag{16}$$

This, however, leads to difficulty in some cases. For example, compare the reactions

$$\underset{\sigma=1}{\mathrm{H}} + \underset{\sigma=2}{\mathrm{H_2}} \to \underset{\sigma=2}{\mathrm{H\cdots H\cdots H}^\ddagger} \to \mathrm{H_2} + \mathrm{H}$$

and

$$\underset{\sigma=1}{\mathrm{H}} + \underset{\sigma=2}{\mathrm{D_2}} \to \underset{\sigma=1}{\mathrm{H\cdots D\cdots D}^\ddagger} \to \mathrm{HD} + \mathrm{D}$$

[1] V. Gold, *Trans. Faraday Soc.*, **60**:739 (1964).

[2] D. M. Bishop and K. J. Laidler, *loc. cit.*

[3] J. N. Murrell and K. J. Laidler, *Trans. Faraday Soc.*, **64**:371 (1968).

Use of symmetry numbers favors the second reaction by a factor of 2. This, however, is intuitively unreasonable; in both cases the atom can approach the molecule from either side, and the same factor is to be expected.   Consider also the following reactions in which a proton is transferred to species S (assumed to have $\sigma = 1$) from the pyramidal ions H_3O^+, H_2DO^+, and HD_2O^+:

$$H_3O^+ + S \rightarrow \overset{\displaystyle \overset{H}{\underset{\displaystyle \quad}{|}}\overset{H}{\diagup}}{O^+}\cdots H\cdots S^\ddagger \rightarrow H_2O + HS^+$$
$$\sigma = 3 \quad\quad \sigma = 1 \quad\quad\quad \sigma = 1$$

$$H_2DO^+ + S \rightarrow \overset{\displaystyle \overset{D}{\underset{\displaystyle \quad}{|}}\overset{H}{\diagup}}{O^+}\cdots H\cdots S^\ddagger \rightarrow DHO + HS^+$$
$$\sigma = 1 \quad\quad \sigma = 1 \quad\quad\quad \sigma = 1$$

$$HD_2O^+ + S \rightarrow \overset{\displaystyle \overset{D}{\underset{\displaystyle \quad}{|}}\overset{D}{\diagup}}{O^+}\cdots H\cdots S^\ddagger \rightarrow D_2O + HS^+$$
$$\sigma = 1 \quad\quad \sigma = 1 \quad\quad\quad \sigma = 1$$

The conventional use of symmetry numbers now predicts statistical factors in the ratios $3:1:1$ for these three reactions; common sense, however, demands $3:2:1$, since this is the ratio of the number of protons being transferred.

These difficulties disappear if symmetry numbers are omitted from the partition functions and the rate equation is multiplied by the appropriate statistical factor; this is evaluated by labeling identical atoms and seeing how many equivalent activated complexes are formed.   Some examples of such statistical factors, designated $l^\ddagger$, are as follows:

$$H + H_2 \rightarrow H\cdots H\cdots H^\ddagger \quad\quad l^\ddagger = 2$$
$$H + D_2 \rightarrow H\cdots D\cdots D^\ddagger \quad\quad l^\ddagger = 2$$
$$H + HD \rightarrow H\cdots H\cdots D^\ddagger \quad\quad l^\ddagger = 1$$
$$H + CH_4 \rightarrow H\cdots H\cdots CH_3^\ddagger \quad\quad l^\ddagger = 4$$
$$H_2DO^+ + S \rightarrow HDO^+\cdots H\cdots S^\ddagger \quad\quad l^\ddagger = 2$$

The justification of this procedure is that the rate of reaction is determined by the number of species *entering* the activated state, which is a point of no return; the question of the number of species returning to the initial state, which enters into the ratio of symmetry numbers, is of no significance as far as rate is concerned.

However, difficulty still remains.   Sometimes a straightforward application of the procedure leads to an inconsistency between the rate constants for forward and reverse reactions; their ratio does not give the correct equilibrium constant.   Consider the reaction

$$H + HX \rightarrow H_2 + X$$

and suppose that we choose as activated complex[1] a linear and symmetrical species $H\cdots X\cdots H^{\ddagger}$.   If we label the hydrogen atoms, we can produce one activated complex from $H + HX$ and one from $H_2 + X$; the rate constants for the reactions in the two directions would therefore be

$$k_1 = \frac{kT}{h} \frac{Q_{\ddagger}^0}{Q_H^0 Q_{HX}^0} e^{-E_1/RT} \tag{17}$$

and

$$k_{-1} = \frac{kT}{h} \frac{Q_{\ddagger}^0}{Q_{H_2}^0 Q_X^0} e^{-E_{-1}/RT} \tag{18}$$

where the Q^0's are the partition functions with the symmetry numbers omitted.   The equilibrium constant for the overall reaction would therefore be

$$K = \frac{k_1}{k_{-1}} = \frac{Q_{H_2}^0 Q_X^0}{Q_H^0 Q_{HX}^0} e^{-(E_1 - E_{-1})/RT} \tag{19}$$

This, however, is incorrect; H_2 has a symmetry number of 2, so that this expression should be divided by 2.   The same difficulty arises if a bent but symmetrical $H\cdots X\cdots H$ species is chosen as the activated complex. The difficulty is avoided if the (more chemically plausible) unsymmetrical $H\cdots H\cdots X$ structure is chosen for the activated state; a statistical factor of 2 is then involved for the back reaction.

Obviously one or both of the rate expressions (17) and (18) must be incorrect.   A further consideration of the problem leads to a significant conclusion about the permissible symmetries for activated complexes. The error lies not in the principle of the statistical-factor method, but in the choice, in the above example, of an activated complex which has too high a degree of symmetry to be an acceptable one for the reaction in question.

This may be seen with reference to a schematic potential-energy surface.   The possible reaction paths for the labeled system leading to the symmetrical activated complex $H\cdots X\cdots H^{\ddagger}$ are as follows:

$$
\begin{array}{c}
H^1 + XH^2 \\
\searrow \\
\qquad\qquad H^{1}\cdots X\cdots H^{2\ddagger} \rightleftharpoons H^1H^2 + X \\
\nearrow \\
H^1X + H^2
\end{array}
$$

---

[1] The fact that this is an unlikely activated complex for this type of reaction is irrelevant to the present argument.

Figure 25 shows a schematic potential-energy surface, and there are three valleys. It can be shown,[1] by an argument based on a Taylor expansion, that three valleys in multidimensional space cannot meet at a *single* col or saddle point, but must meet at three separate activated states, each of which represents the lowest pass between a pair of valleys. These three activated states are shown on the diagram.

It follows that a valid activated complex can only exist at the intersection of *two* valleys. If, therefore, for the H + HX reaction the complex were really of the form $H\cdots X\cdots H^\ddagger$, it would have to be at least slightly unsymmetrical, so that it can be formed in two ways from $H_2 + X$:

$$H^1 + XH^2 \rightleftharpoons H^1\cdots X\cdots H^{2\ddagger}$$

$$\searrow$$
$$\qquad\qquad\qquad H^1H^2 + X$$
$$\nearrow$$

$$H^2 + XH^1 \rightleftharpoons H^2\cdots X\cdots H^{1\ddagger}$$

The expression in Eq. (17) must therefore be multiplied by 2, and the correct equilibrium constant is then obtained.

The important point is that one must always choose an activated complex that will lead to only one set of reactants and to one set of products. If this is done, correct rate equations, consistent with the equilibrium expression, will be obtained. Thus, consider the general reaction

$$A + B \underset{r_1^\ddagger}{\overset{l_1^\ddagger}{\rightleftharpoons}} X^\ddagger \underset{l_{-1}^\ddagger}{\overset{r_{-1}^\ddagger}{\rightleftharpoons}} C + D$$

where the statistical factors are as indicated. The ratio of statistical fac-

[1] J. N. Murrell and K. J. Laidler, *loc. cit.*

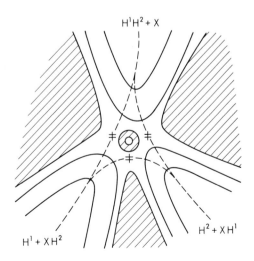

Fig. 25 Schematic potential-energy surface for the reaction H + XH → $H_2X^\ddagger$ → H_2 + X. The true surface is a multidimensional one. Regions of higher energy are shaded; the three cols are denoted by $\ddagger$.

tors must be the ratio of symmetry numbers, i.e.,

$$\frac{\sigma_A \sigma_B}{\sigma_\ddagger} = \frac{l_1^\ddagger}{r_1^\ddagger} \tag{20}$$

and

$$\frac{\sigma_\ddagger}{\sigma_C \sigma_D} = \frac{r_{-1}^\ddagger}{l_{-1}^\ddagger} \tag{21}$$

The rate expressions are

$$k_1 = l_1^\ddagger \frac{\mathbf{k}T}{h} \frac{Q_\ddagger}{Q_A Q_B} e^{-E_1/RT} \tag{22}$$

$$k_{-1} = l_{-1}^\ddagger \frac{\mathbf{k}T}{h} \frac{Q_\ddagger}{Q_C Q_D} e^{-E_{-1}/RT} \tag{23}$$

and the equilibrium constant is

$$K = \frac{l_1^\ddagger}{l_{-1}^\ddagger} \frac{Q_C Q_D}{Q_A Q_B} e^{-(E_1 - E_{-1})/RT} \tag{24}$$

$$= \frac{\sigma_A \sigma_B}{\sigma_C \sigma_D} \frac{Q_C Q_D}{Q_A Q_B} e^{-(E_1 - E_{-1})/RT} \tag{25}$$

provided that the activated complex is chosen so that $r_1^\ddagger = r_{-1}^\ddagger = 1$. This is the correct equilibrium expression.

The conventional procedure of using symmetry numbers does give the correct rate equation provided that the activated complex is such that $r_1^\ddagger = r_{-1}^\ddagger = 1$. However, it seems preferable to formulate rate equations in terms of statistical factors, as in Eq. (22), as it is then easily verified that the activated complex is of an acceptable type.

The rule that a valid activated complex must be of sufficiently low symmetry that $r_1^\ddagger = r_{-1}^\ddagger = 1$ is a useful one in excluding certain structures. A simple example is provided by the exchange reaction between D and H_2O. Symmetrical planar and symmetrical nonplanar activated complexes are both excluded for this reaction, as is shown by the statistical factors given below:

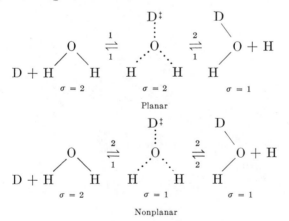

Planar

Nonplanar

A nonsymmetrical planar or nonplanar activated complex is, however, acceptable; e.g.,

$$D + H \quad \underset{\substack{4\\1}}{\overset{\text{}}{\rightleftharpoons}} \quad \underset{\text{Nonplanar}}{O} \quad \underset{\substack{1\\2}}{\overset{\text{}}{\rightleftharpoons}} \quad O + H$$

$$\sigma = 2 \qquad\qquad \sigma = \tfrac{1}{2} \qquad\qquad \sigma = 1$$

Since the activated complex is a mixture of two enantiomers, its symmetry number is taken as $\tfrac{1}{2}$.

Another example is the isomerization of cyclopropane to propylene,

$$\begin{array}{c} CH_2 \\ H_2C \!-\!\!-\!\!-\! CH_2 \end{array} \rightarrow CH_3\!-\!CH = CH_2$$

For this reaction a symmetrical activated complex, with one C—C bond extended, has sometimes been postulated. The statistical factors would then be as follows:

$$\begin{array}{c} CH_2 \\ H_2C\!-\!\!-\!CH_2 \\ \sigma = 6 \end{array} \underset{1}{\overset{3}{\rightleftharpoons}} \begin{array}{c} CH_2^{\ddagger} \\ H_2C\cdots\cdots CH_2 \\ \sigma = 2 \end{array} \underset{3}{\overset{2}{\rightleftharpoons}} CH_3CH\!\!=\!\!CH_2$$

Since the statistical factor from the activated state to $CH_3CH\!=\!CH_2$ is 2, this structure for the activated complex is excluded. It is necessary to postulate a complex that is sufficiently unsymmetrical that it can form propylene in only one way, for example,

$$\begin{array}{c} CH_2 \\ H_2C\!-\!\!-\!CH_2 \\ \sigma = 6 \end{array} \underset{1}{\overset{12}{\rightleftharpoons}} \begin{array}{c} H\cdots CH^{\ddagger} \\ H_2C\cdots CH_2 \\ \sigma = \tfrac{1}{2} \end{array} \underset{6}{\overset{1}{\rightleftharpoons}} CH_3CH\!\!=\!\!CH_2$$

Nonplanar

The activated complex is necessarily nonplanar; a planar complex would be able to produce cyclopropane in two ways. Another example is

$$\sigma = 2 \underset{1}{\overset{4}{\rightleftharpoons}} \underset{\substack{\text{Nonplanar}\\ \sigma = \tfrac{1}{2}}}{\ddagger} \underset{8}{\overset{1}{\rightleftharpoons}} \sigma = 2 \quad + H_2 \quad \sigma = 2$$

A planar complex is again excluded.

FREE-ENERGY SURFACES

So far in this discussion it has been assumed that the activated complexes are those species that exist in the neighborhood of the saddle point of a *potential-energy* surface. It has sometimes been suggested that the activated state should be defined instead with reference to a *free-energy* surface. Various discussions of this problem have been given,[1] and there is no general agreement.

It is not possible to give a firm answer to this question, since the whole concept of the activated complex is an artificial (although enormously useful) one. Strictly speaking, rates should be calculated by performing dynamical calculations with reference to potential-energy surfaces; the motions of systems over the surface would be considered in detail, and the concept of the activated complex would then not arise. However, if as a useful approximation we *are* going to work with activated complexes, it is obviously of some importance to ask how they should best be defined, especially as the two alternative definitions, in terms of potential-energy and free-energy surfaces, sometimes lead to quite different answers.

In considering this question we should first remind ourselves of what the basic assumptions of activated-complex theory are:

1. We define activated complexes in some way and calculate their concentrations with an equilibrium formulation.
2. We multiply this concentration by a frequency ν, given by

$$\nu = \left(\frac{kT}{2\pi m}\right)^{1/2} \frac{1}{\delta} \tag{26}$$

It is to be noted from this second assumption that the motion of the activated complexes is free and spontaneous, with no barrier to overcome. We note also from Eq. (26) that the temperature enters into ν, and this reminds us that we are not concerned with the motion of individual systems over the top of the barrier, but that we are dealing with a statistical distribution of activated complexes.

If we were concerned with the motion of individual molecules over the barrier, the factor determining whether free and spontaneous motion would occur would be the potential energy; the individual species would be free from restraints if X were at the top of a *potential-energy* barrier. However, we are not dealing in this theory with individual molecules, but with an assembly of molecules in a statistical distribution among energy

[1] M. Szwarc, *Chem. Soc. (London) Spec. Pub.*, **16**:25 (1962); K. J. Laidler and J. C. Polanyi, in G. Porter (ed.), "Progress in Reaction Kinetics," pp. 37–39, Pergamon Press, Oxford, 1965.

states. The question of whether or not there is spontaneous motion for such a statistical distribution is determined by the *free energy* and not by the potential energy (at $T = 0$ the two of course become the same, but at other temperatures they are different).

Figure 26 shows schematic potential-energy and free-energy profiles, the highest points on the two being at different positions. If we were concerned with an individual molecule, the question of whether it could move freely over the barrier would certainly depend on whether it possessed energy equal to the maximum of the potential-energy curve. With a statistical distribution of systems, however, the situation is entirely different. A statistical assembly of species having potential energy corresponding to point A would not be able to move spontaneously to the right on the diagram, since such a movement would involve an increase in standard free energy. It is when an assembly of systems is represented by point B that spontaneous motion in both directions is permitted.

An alternative way of looking at the matter is with reference to Fig. 27, which shows sections through the potential-energy surface, normal to the lowest reaction path. At $T = 0$, or for an individual molecule, there is spontaneous motion from A to B, since the zero-point level for A is higher than for B. For a statistical assembly of molecules at a higher temperature, however, the situation might be reversed; the systems may distribute themselves over the energy levels in such a way that there will be spontaneous motion from B to A. Position B corresponds to a tighter structure than A; ν is larger, the energy levels more separated, and the entropy lower. The free energy corresponding to B therefore may be higher than at A, and the spontaneous motion is from B to A.

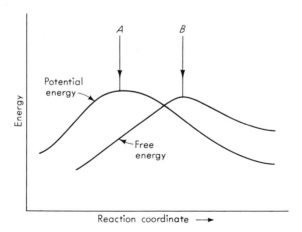

Fig. 26 Schematic plots of potential energy and free energy against the reaction coordinate.

When we look at matters from this point of view, we seem to be forced to the conclusion that if we think in terms of activated complexes, we should logically regard them as the species of maximum *free energy* along the reaction path. It is to be admitted, however, that in view of the artificiality of the activated complex this is not a compelling argument; other ways of looking at the matter can lead to different conclusions.

We may now ask how one would go about constructing a free-energy surface, a task that has apparently never been attempted. The problem is certainly not entirely straightforward. Whereas a potential-energy surface applies to individual molecules, an entropy or a free-energy surface would apply to an assembly of reactants and activated complexes in a statistical distribution. The form of an entropy surface or a free-energy surface would vary with temperature, unlike that of a potential-energy surface.

Two alternative procedures might be used for constructing free-energy surfaces; both of them start with a potential-energy surface.

1. The lowest reaction path could be constructed on the surface. At each point along the reaction path we could calculate the vibrational frequencies corresponding to the potential-energy variations in directions normal to the reaction path. From these vibrational frequencies, entropies and hence free energies could be calculated. In this way we would obtain a free-energy profile along the minimal reaction path, and this would locate the activated complex.

2. If a complete dynamical treatment has been made of the motion over the potential-energy surface, free energies at any point on the surface can be calculated in the following way. If a swarm of systems is traveling over the surface, the relative probability of each configuration can be obtained by the density of the swarm at the corresponding point. From this probability the entropy can be calculated; this combined with the potential energy gives the free energy.

There are several situations where it seems to be more fruitful to think in terms of free-energy rather than potential-energy surfaces. One of these is in connection with unimolecular decompositions and the reverse

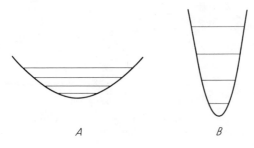

Fig. 27 Sections through a potential-energy surface, normal to the lowest reaction path; *A* is at the col, *B* at some other position.

A   *B*

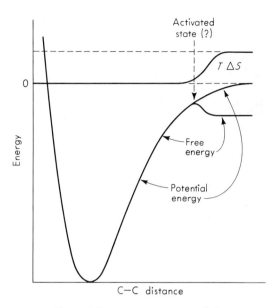

Fig. 28 Potential-energy, entropy and free-energy profiles for the dissociation of C_2H_6 into $2CH_3$. If the entropy variation is as shown, the free energy can pass through a maximum as the C—C bond is stretched.

recombinations of free radicals.[1] On a potential-energy surface the activated state is not unambiguously defined, since the potential-energy does not go through a maximum but rises to a limiting value as dissociation proceeds. This is shown for the dissociation of C_2H_6 in the potential-energy profile shown in Fig. 28. However, as dissociation proceeds, there is an increase in entropy, owing to the loosening of vibrations and the development of internal rotations; it is therefore possible that the free energy passes through a maximum. It is to be noted that according to this point of view it is possible for the activation energy of a dissociation to be less than the endothermicity and for the association process to have a negative energy of activation. This matter is capable of experimental test; so far no clear-cut example of this behavior has been reported.

[1] C. Steel and K. J. Laidler, *J. Chem. Phys.*, **34**:1827 (1961).

5

Bimolecular and Trimolecular Reactions

In the present chapter a brief and general account will be given of some of the work that has been done on the application of activated-complex theory to reactions in the gas phase. Since the main object of this book is to review the theories of reaction rates, their application will not be dealt with in any detail, but references will be given to publications where a much more thorough treatment is given.

Unimolecular reactions present a special problem and are treated separately in the next chapter. The reason they are different from ordinary bimolecular and trimolecular reactions is that they involve an intermediate of fairly long life, so that problems of inter- and intramolecular energy transfer become of primary importance. Activated-complex theory is not concerned with these problems, since it focuses attention on the equilibrium states of activated complexes and provides no interpretation of how these complexes are formed from the reactants.

Bimolecular and trimolecular gas reactions can be divided into two classes, those that are the reverse of unimolecular decompositions and those that are not. Examples of reactions in the first class are the combinations of atoms and free radicals, such as

$$I + I + M \rightarrow I_2 + M$$
$$CH_3 + CH_3 + M \rightarrow C_2H_6 + M$$

where M is a third body. These reactions do not occur by a *direct* mechanism (see page 45); instead, an intermediate (such as I_2^* or IM^*) is formed, and this undergoes a subsequent reaction. Such reactions must therefore not be treated by a simple application of activated-complex theory.

Bimolecular and trimolecular reactions that are not the reverse of unimolecular decompositions generally occur by a *direct* mechanism and may therefore be treated by activated-complex theory. As has been indicated (see also page 213), the assumptions of activated-complex theory are only expected to break down for very fast reactions, and it will be seen in the present chapter that for ordinary reactions the theory has been quite successful.

REACTIONS BETWEEN ATOMS

Although atom-combination reactions are strictly speaking outside the scope of the present chapter, it is nevertheless instructive to make a formal application of activated-complex theory to them. By so doing we are led to the interesting result that activated-complex theory is equivalent to the hard-sphere theory of collisions when it is applied to a reaction between structureless particles.

For a reaction between two atoms A and B the activated complex is the diatomic species $A \cdots B^\ddagger$. It has three degrees of translational freedom and two of rotational. If it were a normal diatomic molecule, it would have one degree of vibrational freedom, but as it is an activated complex this mode of vibration corresponds to passage through the activated state and is therefore omitted. The partition function $Q_\ddagger$ is therefore

$$Q_\ddagger = \frac{[2\pi(m_A + m_B)\mathbf{k}T]^{3/2}}{h^3} \frac{8\pi^2 I^\ddagger \mathbf{k}T}{h^2} \tag{1}$$

where m_A and m_B are the atomic masses, and $I^\ddagger$ is the moment of inertia. If d_{AB} is the distance between the centers of the atoms in the activated complex, the moment of inertia is

$$I^\ddagger = d_{AB}{}^2 \frac{m_A m_B}{m_A + m_B} \tag{2}$$

The partition functions for the reactants are

$$Q_A = \frac{(2\pi m_A \mathbf{k}T)^{3/2}}{h^3} \tag{3}$$

and

$$Q_B = \frac{(2\pi m_B \mathbf{k} T)^{3/2}}{h^3} \tag{4}$$

With a transmission coefficient of unity the rate constant is therefore

$$k = \frac{\mathbf{k} T}{h} \frac{Q_\ddagger}{Q_A Q_B} e^{-E_0/RT} \tag{5}$$

$$= d_{AB}{}^2 \left(8\pi \mathbf{k} T \frac{m_A + m_B}{m_A m_B} \right)^{1/2} e^{-E_0/RT} \tag{6}$$

This expression is equivalent to that given by Trautz[1] and Lewis[2] on the basis of a collision theory in which the reactants are treated as hard spheres having radii r_A and r_B such that $r_A + r_B$ is equal to d_{AB}. This result is of particular significance, since it indicates that the hypothesis of equilibrium between initial and activated states is implicit in these simple collision treatments.

BIMOLECULAR REACTIONS BETWEEN MOLECULES

In treating reactions between more complicated molecules it is useful to begin with a grossly oversimplified treatment which, as long as it is not taken too seriously, can be very instructive. The approximation will be made of regarding the partition function for each type of energy as consisting of equal factors for each degree of freedom. This procedure is to some extent justified by the orders of magnitude shown in Table 3, page 47, for the different types of motion; thus, partition functions for a single degree of translational freedom are generally of the order of 10^8, while those for rotation are often of the order of 10. Those for vibration are close to unity for ordinary molecules except at elevated temperatures; however, it should be noted that vibrational frequencies in activated complexes are frequently abnormally low, so that the partition functions may be much greater than unity.

With the contributions of single translational, rotational, and vibrational degrees of freedom written as q_T, q_R, and q_v, the total partition function for a molecule may be expressed as

$$Q = q_T{}^t q_R{}^r q_v{}^v \tag{7}$$

where t, r, and v are the numbers of degrees of freedom. For the case already considered, reaction between two atoms,

$$Q_A = q_T{}^3 \qquad Q_B = q_T{}^3 \qquad Q_\ddagger = q_T{}^3 q_R{}^2 \tag{8}$$

[1] M. Trautz, Z. Anorg. Allgem. Chem., **96**:1 (1916).
[2] W. C. McC. Lewis, J. Chem. Soc., **113**:471 (1918).

so that the rate constant is given by

$$k_a = \frac{\mathbf{k}T}{h} \frac{Q_{\ddagger}}{Q_A Q_B} e^{-E_0/RT} \tag{9}$$

$$= \frac{\mathbf{k}T}{h} \frac{q_R{}^2}{q_T{}^3} e^{-E_0/RT} \tag{10}$$

In the general case, with A and B considered nonlinear molecules containing N_A and N_B atoms, respectively,

$$Q_A = q_T{}^3 q_R{}^3 q_V{}^{3N_A-6} \tag{11}$$

$$Q_B = q_T{}^3 q_R{}^3 q_V{}^{3N_B-6} \tag{12}$$

$$Q_{\ddagger} = q_T{}^3 q_R{}^3 q_V{}^{3(N_A+N_B)-7} \tag{13}$$

The rate constant is now

$$k_m = \frac{\mathbf{k}T}{h} \frac{q_V{}^5}{q_T{}^3 q_R{}^3} e^{-E_0/RT} \tag{14}$$

The preexponential factor in Eq. (10) is that given by simple collision theory; that in (14) differs from it by a factor of

$$\left(\frac{q_V}{q_R}\right)^5$$

In general, q_V is a good deal less than q_R, so that abnormally small frequency factors are expected for reactions between complex molecules. With less complex reacting species the discrepancy is not so great; thus in the case of reaction between an atom and a diatomic molecule, with the formation of a nonlinear activated complex, the factor is q_V/q_R.

This type of argument is useful but must not be taken too seriously. Thus Johnston[1] has pointed out that if one examines the data for the reactions of atoms (H, Cl, or Br) with a series of molecules of increasing complexity (H_2, CH_4, C_2H_6, etc.), there is no decrease of preexponential factor with the number of atoms. It is too much to expect of the crude treatment that there should be such a decrease. If, on the other hand, one compares the reactions of atoms (for example, H) with organic molecules and the reactions of free radicals (for example, CH_3) with the same molecules, one finds the preexponential factors to be of the order of 10^{13} to 10^{14} cc mole^{-1} sec^{-1} in the former case and 10^{11} in the latter (see Table 5). This large difference can realistically be ascribed to the difference in complexity of the reactants in the two cases.

Trends of this kind can also be interpreted in terms of entropy effects. For a reaction between atoms there are no rotational contribu-

[1] H. S. Johnston, "Gas Phase Reaction Rate Theory," p. 221, The Ronald Press Company, New York, 1966.

tions, but in the activated state there are two degrees of rotational freedom.   For a reaction between complex molecules there are initially three
degrees of rotational freedom for each reactant, but there are only three
altogether in the activated complex.   There is thus a net loss of three
degrees of rotational freedom, so that compared with the atomic (simple
collision-theory) case there is a relative loss of five degrees of rotational
freedom.   To compensate for this there is a gain of five degrees of vibrational freedom.   However, the entropy associated with vibration is usually much less than that associated with rotation.   Entropies of activation
are therefore expected to be more negative, the more complex the reactants are.

Arguments of this kind are instructive in that they show that
activated-complex theory does correctly interpret the main trends that
are observed in frequency factors.   No such interpretation is given in
collision theory which, if anything, predicts a slight increase in frequency
factor as the complexity of the reactants increases, owing to the increase
in collision diameter.

A number of much more detailed tests of activated-complex theory
have been made.   They can be roughly divided into two classes:

1. Order-of-magnitude calculations of frequency factors, estimates being
   made of vibrational frequencies in the activated state
2. Calculations of the rate constants of a series of reactions in which
   there has been isotopic substitution, for example, of H by D

Order-of-magnitude calculations of frequency factors have been
made for a variety of reactions of different types, and a number of
results were given by Glasstone, Laidler, and Eyring.[1]   More recent tests
have been made by Hill,[2] Bywater and Roberts,[3] Polanyi,[4] Knox and
Trotman-Dickenson,[5] Herschbach, Johnston, Pitzer, and Powell,[6] Wilson
and Johnston,[7] Johnston and Goldfinger,[8] and Goldfinger and Martens[9]

[1] S. Glasstone, K. J. Laidler, and H. Eyring, "The Theory of Rate Processes," Mc-
Graw-Hill Book Company, New York, 1941.

[2] T. L. Hill, *J. Chem. Phys.*, **17**:503 (1949).

[3] S. Bywater and R. Roberts, *Can. J. Chem.*, **30**:773 (1952).

[4] J. C. Polanyi, *J. Chem. Phys.*, **23**:1505 (1955); **24**:493 (1956).

[5] J. H. Knox and A. F. Trotman-Dickenson, *J. Phys. Chem.*, **60**:1367 (1956).

[6] D. R. Herschbach, H. S. Johnston, K. S. Pitzer, and R. E. Powell, *J. Chem. Phys.*,
25:736 (1956).

[7] D. J. Wilson and H. S. Johnston, *J. Am. Chem. Soc.*, **79**:29 (1957).

[8] H. S. Johnston and P. Goldfinger, *J. Chem. Phys.*, **37**:700 (1962).

[9] P. Goldfinger and G. Martens, *Trans. Faraday Soc.*, **57**:2220 (1961).

on a variety of gas-phase reactions, some involving atoms and free radicals. In all cases the calculations led to frequency factors that were within an order of magnitude of the experimental values. As examples, some of the results of the calculations of Herschbach, Johnston, Pitzer, and Powell are given in Table 4. It is to be seen that the activated-complex-theory calculations of the frequency factors are reasonably close to the experimental values. The simple collision theory, on the other hand, leads to A values of close to 10^{14} cc mole^{-1} sec^{-1}, which is significantly too high in all cases. Another comparison, for a series of atom and free-radical reactions, is shown in Table 5. Again the agreement is quite satisfactory. The results in this table are of special interest in showing the fall in frequency factor as the reacting molecules become more complex. The reactions of hydrogen atoms have frequency factors that are sometimes described as "normal"; this implies that they are not far from the collision-theory estimate of 10^{13} to 10^{14}. The reactions of methyl radicals, on the other hand, have values of the order of 10^{11}, a fact that has no interpretation in simple collision theory but is readily interpreted by activated-complex theory in the loss of rotational freedom in the activated state.

Table 4   Kinetic parameters for some bimolecular reactions

Reaction	Activation energy, kcal/mole	Logarithm of frequency factor, cc mole^{-1} sec^{-1}			Reference
		Observed	Calculated by activated-complex theory	Calculated by simple collision theory	
$NO + O_3 \rightarrow NO_2 + O_2$	2.5	11.9	11.6	13.7	a
$NO + O_3 \rightarrow NO_3 + O$	7.0	12.8	11.1	13.8	b
$NO_2 + F_2 \rightarrow NO_2F + F$	10.4	12.2	11.1	13.8	c
$NO_2 + CO \rightarrow NO + CO_2$	31.6	13.1	12.8	13.6	d
$2NO_2 \rightarrow 2NO + O_2$	26.6	12.3	12.7	13.6	e
$NO + NO_2Cl \rightarrow NOCl + NO_2$	6.9	11.9	11.9	13.9	f
$2NOCl \rightarrow 2NO + Cl_2$	24.5	13.0	11.6	13.8	g
$NO + Cl_2 \rightarrow NOCl + Cl$	20.3	12.6	12.1	14.0	h
$F_2 + ClO_2 \rightarrow FClO_2 + F$	8.5	10.5	10.9	13.7	i
$2ClO \rightarrow Cl_2 + O_2$	0	10.8	10.0	13.4	j

[a] H. S. Johnston and H. J. Crosby, *J. Chem. Phys.*, **22**:689 (1954).

[b] H. S. Johnston and D. M. Yost, *J. Chem. Phys.*, **17**:386 (1949).

[c] R. L. Perrine and H. S. Johnston, *J. Chem. Phys.*, **21**:2200 (1953).

[d] H. S. Johnston, W. A. Bonner, and D. J. Wilson, *J. Chem. Phys.*, **26**:1002 (1957).

[e] M. Bodenstein and H. Ramstetter, *Z. Physik. Chem (Leipzig)*, **100**:106 (1922).

[f] E. C. Freiling, H. S. Johnston, and R. A. Ogg, *J. Chem. Phys.*, **20**:327 (1952).

[g] G. Waddington and R. C. Tolman, *J. Am. Chem. Soc.*, **57**:689 (1935).

[h] P. G. Ashmore and J. Chanmugan, *Trans. Faraday Soc.*, **49**:270 (1953).

[i] P. J. Aynoneno, J. E. Sicre, and H. J. Schumacher, *J. Chem. Phys.*, **22**:756, (1954).

[j] G. Porter and F. J. Wright, *Discussions Faraday Soc.*, **14**:23 (1953).

Table 5   Activation energies and frequency factors for some bimolecular reactions involving atoms and free radicals

Reaction	Activation energy, kcal/mole	log A, A, cc mole^{-1}sec^{-1}	
		Observed[a]	Calculated by activated-complex theory
H + H$_2$ → H$_2$ + H	8.8	14.0	13.7,[b] 13.8,[c] 13.7[d]
Br + H$_2$ → HBr + H	17.6	13.5	14.1[d]
H + CH$_4$ → H$_2$ + CH$_3$	12	13	13.3[c]
H + C$_2$H$_6$ → H$_2$ + C$_2$H$_5$	6.8	12.5	13.1[b]
CH$_3$ + H$_2$ → CH$_4$ + H	10.0	12.3	12.0,[b] 12.4,[c] 12.0[d]
CD$_3$ + CH$_4$ → CD$_3$H + CD$_3$	14.0	11	11.3,[d] 10.9[b] (for CH$_3$)
CH$_3$ + C$_2$H$_6$ → CH$_4$ + C$_2$H$_5$	11.2	10.8	11.0[b]
CD$_3$ + C$_2$H$_6$ → CD$_3$H + C$_2$H$_5$	10.4	11.3	11.3[d]
CH$_3$ + iso-C$_4$H$_{10}$ → CH$_4$ + C$_4$H$_9$	7.6	10	9.8[b]
CH$_3$ + n-C$_5$H$_{12}$ → CH$_4$ + C$_5$H$_{11}$	8.1	11.0	
CH$_3$ + CH$_3$COCH$_3$ → CH$_4$ + CH$_2$COCH$_3$	9.7	11.6	11[e]
CD$_3$ + CD$_3$COCD$_3$ → CD$_4$ + CD$_2$COCD$_3$	11.3	11.8	
CD$_3$ + C$_6$H$_6$ → CD$_3$H + C$_6$H$_5$	9.2	10.4	
CH$_3$ + C$_6$H$_5$CH$_3$ → CH$_4$ + C$_6$H$_5$CH$_2$	7	10	
CF$_3$ + CH$_4$ → CF$_3$H + CH$_3$	9.5	11.0	
CF$_3$ + C$_2$H$_6$ → CF$_3$H + C$_2$H$_5$	7.7	11.4	

[a] The experimental values are discussed, with references, by E. W. R. Steacie, "Atomic and Free Radical Reactions," Reinhold Publishing Corporation, New York, 1954, and by K. O. Kutschke and E. W. R. Steacie in "Vistas in Free Radical Chemistry," p. 162, Pergamon Press, Oxford, 1959. There is some doubt on the experimental side about the results for the reaction between H and CH$_4$; the values given represent the author's opinion on the most reliable figures. However, work by J. W. S. Jamieson and G. R. Brown, *Can. J. Chem.*, **42**:1638 (1964), leads to much lower values of E and log A (7.4 kcal and 12).

[b] S. Bywater and R. Roberts, *Can. J. Chem.*, **30**:773 (1952).

[c] J. C. Polanyi, *J. Chem. Phys.*, **23**:1505 (1955); **24**:493 (1956).

[d] D. J. Wilson and H. S. Johnston, *J. Am. Chem. Soc.*, **79**:29 (1957).

[e] T. L. Hill, *J. Chem. Phys.*, **17**:503 (1949).

KINETIC ISOTOPE EFFECTS

It is outside the scope of the present book to give a systematic treatment of the general theory of kinetic isotope effects. A number of reviews have been published on this subject.[1] The present section merely outlines the main factors on which the isotope effects depend and indicates some of the comparisons that have been made with experiment on the basis of activated-complex theory.

The theoretical treatment of kinetic isotope effects is very complicated and cannot easily be applied in detail to individual reactions. This arises because, when an isotopic substitution is made, a number of rela-

[1] See, for example, L. Melander, "Isotope Effects on Reaction Rates," The Ronald Press Company, New York 1960; J. W. Bigeleisen and M. Wolfsberg, in I. Prigogine (ed.), "Advances in Chemical Physics," vol. 1, Interscience Publishers, New York, 1958; a much briefer account is given by K. J. Laidler, "Chemical Kinetics," 2d ed., pp. 93–98, McGraw-Hill Book Company, New York, 1965.

tively small effects are involved, all of which must be treated accurately for it to be possible for the overall effect to be predicted.

When an atom is replaced by an isotope, there is no change in the potential-energy surface for any reaction that it might undergo. The reason that there is a change in rate is that there is a change in the average vibrational energy of the molecule and of the activated complex. This may be understood with reference to the species H_2, HD, and D_2. The potential-energy curves are identical for all three species, but the zero-point levels are different; their values, relative to the minimum in the curve, are 6.18, 5.36, and 4.39 kcal, respectively. At relatively low temperatures the molecules reside largely at their zero-point levels. It therefore follows that the molecule H_2 requires less energy (103.22 kcal) to reach its dissociated state than does D_2, for which the energy needed is 105.02 kcal. Thus a reaction in which the dissociation of H_2 is involved will, if no other factors are important, occur more rapidly than one involving HD or D_2.

With more complex molecules the situation is qualitatively similar. Consider, for example, a reaction involving a molecule containing a C—H bond. This molecule executes a complicated type of vibration, and there will be $3N - 6$ normal modes, where N is the number of atoms in the molecule. To a very good approximation some of these frequencies can be regarded as associated with individual bonds; most molecules containing a C—H bond, for example, have a vibrational frequency of 2,900 to 3,000 cm^{-1}, and this is attributed to the presence of the C—H bond. Replacement of the H atom by D changes this vibrational frequency to between 2,000 and 2,100 cm^{-1}. The ratio of the two frequencies is approximately equal to the square root of the ratio of the masses of H and D, and this fact supports the conclusion that these frequencies are associated with the individual bonds.

The situation regarding the reaction of a molecule containing a C—H or C—D bond is therefore very similar to that regarding the reactions of H_2, HD, and D_2. An ordinary C—H bond vibration has a zero-point energy that is greater by about 1.2 kcal than that of the C—D bond. The average vibrational energy of the molecule containing the C—H bond will therefore be higher, by approximately this amount, than that of the corresponding molecule in which H has been replaced by D. If, therefore, the two molecules are involved in a reaction for which the C—H or C—D bond is completely or largely broken in the activated complex, the activation energy of the reaction will be higher for the compound in which H has been replaced by D; the rate will therefore be greater for the light molecule.

If, on the other hand, the bond involving the H or D atoms is far from completely broken in the activated state, the situation is different.

Figure 29 shows the energy relationships in the normal and activated states. The curve corresponding to the normal state is the potential-energy curve for one particular vibration, such as one involving a C—H or C—D bond. The zero-point levels are shown, with the level for C—H higher than that for C—D. The upper curve corresponds to one of the vibrations in the activated state. If the C—H or C—D bond is completely broken in the activated state, the zero-point levels are obviously the same for the light and heavy complexes and the rate will be greater for the light molecule, as discussed above. If the C—H or C—D bond does not change at all when the activated complex is formed, the difference between the zero-point levels will be the same as in the initial state; the energy of activation will then be the same for both compounds, and the isotope effect will be small. If the C—H or C—D bond is weakened, there will be a smaller isotope effect than when the bond is completely broken.

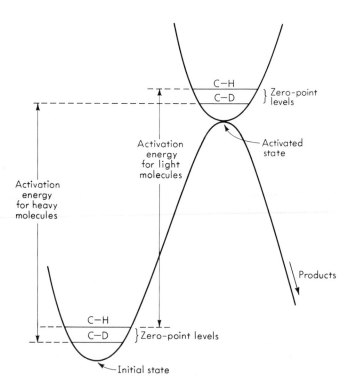

Fig. 29 Potential-energy diagram for a reaction, showing the effect of changing from one isotope to another. For the case illustrated, the C—H or C—D bond remains practically the same in the activated state, so that the isotope effect is very small.

It is possible for the bond involving a hydrogen atom to become stronger in the activated state; this can occur, for example, if a hydrogen atom is transferred from one atom to another. Replacement of H by D will then lead to a greater decrease in energy in the activated state than in the initial state, and the activation energy will therefore be less for the heavy molecule. Under these circumstances there will be an inverse or reverse isotope effect, the heavy molecule reacting more rapidly than the light one.

Reactions of the type H + H$_2$ During recent years a considerable amount of experimental and theoretical work has been done on the reactions

1. $H + H_2 \rightarrow H_2 + H$
2. $D + D_2 \rightarrow D_2 + D$
3. $D + H_2 \rightarrow DH + H$
4. $H + D_2 \rightarrow HD + D$

In spite of this, the situation remains very confused on both the theoretical and experimental sides; it is not entirely clear, for example, whether tunneling occurs in any of these systems.

The rates of all four reactions, over a range of temperatures, have recently been measured by Le Roy and coworkers,[1] who used a fast-flow system of mixing the atoms with the molecules. Their results are shown as Arrhenius plots in Fig. 30. The plots for reactions 1 and 3, in which there is the transfer of an H atom, show clear evidence of curvature at the lower temperatures, a result that tends to suggest that tunneling is involved. It is unfortunate that the work with the other two reactions, 2 and 4, was not done down to the same low temperatures. These reactions involve the transfer of D atoms, so that tunneling is less likely; linearity in the Arrhenius plots would therefore be expected, and this would have provided a useful comparison for the results on reactions 1 and 3.

Weston[2] has analyzed these results for reactions 1 and 4 in terms of various potential-energy surfaces. Figure 31 shows his plot of the rate-constant ratio k_1/k_4 against $1/T$, and the curves are the predictions based on various models. The semiempirical LEP potential-energy surface gives much too small a value for the ratio, even when the tunneling correction is included. The barrier corresponding to this surface is rather flat,

[1] W. R. Schulz and D. J. Le Roy, *Can. J. Chem.*, **42**:2480 (1964); *J. Chem. Phys.*, **42**:3869 (1965); B. A. Ridley, W. R. Schulz, and D. J. Le Roy, *J. Chem. Phys.* **44**:3344 (1966); D. J. Le Roy, B. A. Ridley, and K. A. Quickert, *Discussions Faraday Soc.*, **44**:92 (1967).

[2] R. E. Weston, *J. Chem. Phys.* **31**:892 (1959); D. Rapp and R. E. Weston, *J. Chem. Phys.*, **36**:2807 (1962); for a review, see R. E. Weston, *Science*, **158**:332 (1967); see also I. Shavitt, *J. Chem. Phys.*, **31**:1359 (1959).

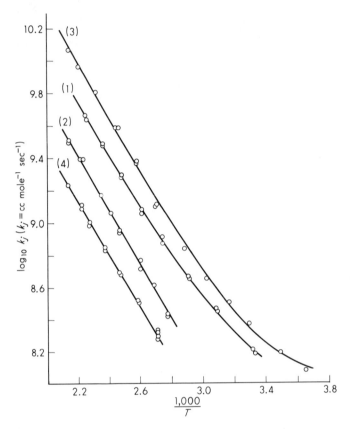

Fig. 30 Arrhenius plots for the results of Le Roy and coworkers on reactions of the type H + H₂.

and the tunneling correction is therefore small. This lack of agreement can be attributed to the fact that the LEP surface, since it has a rather extensive basin, gives an activated complex in which the H_2 or D_2 molecule is only slightly perturbed by the approaching H atom; the force constants for the activated state are therefore similar to those of the reactant molecule, so that the predicted rate-constant ratio is small.

Calculations which ignore the tunneling correction and are based on the semiempirical Sato surface (see page 30), the BEBO surface (page 32), and the quantum-mechanical surface of Boys and Shavitt (page 22) give results that are in quite satisfactory agreement with experiment. Unfortunately, however, as seen from Fig. 31, the agreement is very much worse if the tunneling correction is included. This arises from the rather thin barriers that are given by these treatments. At the present stage it is impossible to say whether these shortcomings are due to faults in the potential-energy surfaces or to the theory of tunneling.[1]

These difficulties are unfortunately compounded by discrepancies on the experimental side.   Westenberg and de Haas[2] have more recently

[1] D. J. Le Roy, B. A. Ridley, and K. A. Quickert [*Discussions Faraday Soc.*, **44**:92 (1967)] have discussed their results on the basis of a modified version of activated-complex theory and have concluded that it is not necessary to invoke the concept of tunneling; however, the validity of their treatment is in serious doubt [see K. J. Laidler, *Discussions Faraday Soc.*, **44**:172 (1967)].

[2] A. A. Westenberg and N. de Haas, *J. Chem. Phys.*, **47**:1393 (1967).

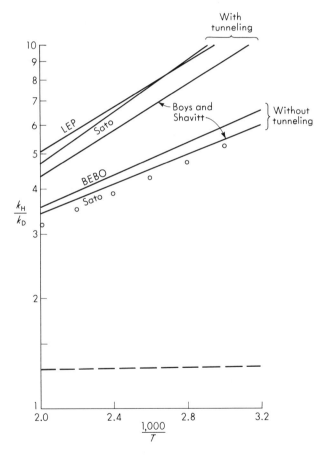

Fig. 31   The ratio of rate constants for the reactions $H + H_2$ and $H + D_2$, as a function of temperature.   The points are the experimental values by Schulz and Le Roy; the lines are theoretical predictions on the basis of various models.

made an experimental study of the reactions

3. $D + H_2 \rightarrow DH + H$

and

4. $H + D_2 \rightarrow HD + D$

They used a fast-flow system and measured the atom concentrations with an electron-spin resonance technique. Their temperature range, 250 to 750°K for reaction 3 and 300 to 750°K for reaction 4, was a good deal wider than that employed by Le Roy and coworkers (for example, 360 to 470°K for reaction 2). Their results are shown as an Arrhenius plot in Fig. 32, on which the results of Le Roy et al. are also shown. There is seen to be a significant difference between the results of the two sets of workers, particularly as far as the degree of curvature is concerned. The results of Westenberg and De Haas show more curvature than those of Le Roy et al. for the D + H₂ reaction. There also appears to be rather more curvature at lower temperatures for the H + D₂ system than for

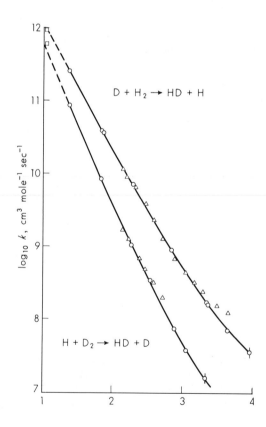

Fig. 32 Arrhenius plots for the results of Westenberg and de Haas (triangles) and of Le Roy et al. (circles) on the reactions H + D₂ and D + H₂.

the $D + H_2$ system; this, of course, is exactly the opposite of what is expected on the basis of tunneling.   Westenberg and de Haas have made calculations based on activated-complex theory and with the use of a Sato potential-energy surface.   Their results are similar to those of Weston, and they obtain good agreement with the experimental values provided that the tunneling correction is omitted.   Inclusion of the tunneling correction leads to much less satisfactory agreement.

More recently Shavitt[1] has made a careful reappraisal of the experimental data for these reactions and has made calculations based on the Shavitt-Stevens-Minn-Karplus potential-energy surface (see page 23), using various procedures for estimating tunneling.   He concludes that the results of Westenberg and de Haas are the most reliable and finds that his calculations are more consistent with these results than with those of Le Roy and coworkers.   The Shavitt-Stevens-Minn-Karplus calculations led to a potential-energy barrier of 11.0 kcal, which is almost certainly too high; Shavitt concludes that the best agreement with experiment is obtained by scaling the surface down by 11 percent, so that the barrier height is 9.8 kcal/mole.   This is somewhat larger than given by earlier estimates from experiment, which had tended to lead to 7 to 9 kcal; as has been emphasized, the relationship between the experimental activation energy and the barrier height depends upon the theory employed. Since Shavitt brings in rather more tunneling than previous calculations, his barrier will have to be higher to accommodate the experimental results.

The treatment of tunneling is a matter of some difficulty, and Shavitt finally selected a barrier of the Eckart form (see page 60) which gave a good fit to the true barrier in its upper regions, even though it did not coincide at the bottom of the barrier (i.e., it corresponded to the wrong barrier height).   On the basis of this he was able to obtain excellent agreement with experiment for the $H + H_2$, $H + D_2$, $D + H_2$, and $D + D_2$ reactions.

More experimental and theoretical work clearly needs to be done on these systems.

Reactions of the type $Cl + H_2$ Rate-constant ratios have been measured by using an isotopic-competition method for the reactions

1. $Cl + H_2 \rightarrow HCl + H$
2. $Cl + D_2 \rightarrow DCl + D$
3. $Cl + HD \rightarrow HCl + D$
4. $Cl + DH \rightarrow DCl + H$

[1] I. Shavitt, *J. Chem. Phys.*, **49**:4048 (1968).

Here the difficulty is in arriving at a reliable potential-energy surface for this system, in which p orbitals are involved. Figure 33 shows an Arrhenius plot for the ratio of reactions 1 and 2 and also predictions based on LEP, Sato, and BEBO surfaces.[1] Again the LEP surface leads to too small an isotope effect. The reason for this is the opposite of that for the $H + H_2$ systems; here the activated complex is too like a product molecule, and again there is considerable cancellation of the isotopic differences between the reactants and activated complexes. The Sato surface, including tunneling (which is quite considerable for this surface), leads to predictions in satisfactory agreement with experiment. Predictions based on the BEBO procedure are too low by 20 to 40 percent.

An extensive isotope study on this reaction was made by Persky and Klein,[2] whose results together with some other results[3] are plotted in Fig. 34. The lines show the predictions obtained with a BEBO surface, including tunneling, and equally good agreement was obtained by using a Sato surface with tunneling.

[1] J. Bigeleisen, F. S. Klein, R. E. Weston, and M. Wolfsberg, *J. Chem. Phys.*, **30**:1340 (1959); G. Chiltz, R. Eckling, P. Goldfinger, G. Huybrechts, H. S. Johnston, L. Meyers, and G. Verbeke, *J. Chem. Phys.*, **38**:1053 (1963); A. Persky and F. S. Klein, *J. Chem. Phys.*, **44**:3617 (1966).

[2] *Ibid.*

[3] Chiltz et al., *loc. cit.*; Bigeleisen et al., *loc. cit.*; W. M. Jones, *J. Chem. Phys.*, **19**:78 (1951).

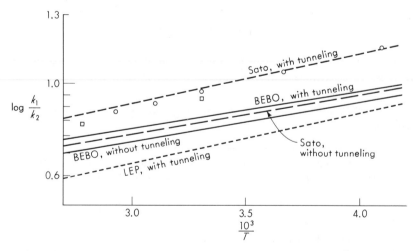

Fig. 33 Arrhenius plot of the ratio k_1/k_2 for the reactions $Cl + H_2$ and $Cl + D_2$. The points are experimental; the lines show the predictions of various theoretical treatments.

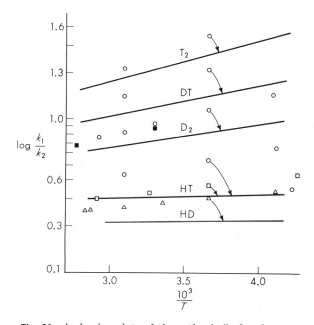

Fig. 34 Arrhenius plots of the ratios k_1/k_2 for the reaction of chlorine atoms with $H_2(k_1)$ and with the other species indicated (k_2). The points are experimental: circles show the results of Persky and Klein; solid squares, Chiltz et al.; hollow squares, Jones; triangles, Bigeleisen et al. The lines are ratios calculated by the BEBO method, with tunneling included.

Reactions of the type Br + H$_2$ Similar studies have been made for the reactions of Br atoms with the hydrogen isotopes.[1] For these systems the activation energy ($\sim$17.6 kcal) is close to the endothermicity ($\sim$16.7 kcal). The theoretical predictions are now in very poor agreement with experiment, giving isotopic ratios that are much too low for all of the surfaces used. The difficulty here may be that the product molecule must be formed with at least its zero-point energy (3.8 kcal), and the effect of this is to force the system to cross the barrier at a level higher than the zero-point energy in the activated state; the assumption of an equilibrium concentration of activated complexes may therefore lead to error. Alternatively, the potential-energy surfaces employed may be far from the truth.

Reactions of the type CF$_3$ + CH$_3$D A number of other hydrogen and deuterium abstractions have been considered by Johnston and his

[1] R. B. Timmons and R. E. Weston, *J. Chem. Phys.*, **41**:1654 (1964).

coworkers.[1]   An interesting example is provided by a comparison of the rate constants for the processes

1. $CF_3 + CHD_3 \rightarrow CF_3H + CD_3$
2. $CF_3 + CHD_3 \rightarrow CF_3D + CHD_2$

This is an *intra*molecular isotope effect, both isotopic species being in the same molecule.   The difference in rate constants therefore depends entirely on the characteristics of the two alternative activated complexes formed, which depend on whether an H or a D atom is abstracted.

Sharp and Johnston[2] treated this problem on the basis of Sato potential-energy surfaces.   In one set of calculations they made the assumption that the energies depend only on the three atoms at the reaction center; in other words, the triatomic system $C\cdots H\cdots C$ was considered, the activated complex being a symmetric linear configuration. They made a whole series of calculations on the basis of models of various complexities, ranging from the triatomic model to the complete nine-atom system.   The various models all give somewhat similar results.   Without tunneling corrections they all predict ratios of rates that are in good agreement with experiment at high temperatures but diverge from the experimental values as the temperature is lowered.   This is shown in Fig. 35, which also shows that the agreement with experiment is greatly improved if tunneling corrections are included.   The evidence for tunneling is not, however, completely unequivocal, since a different choice of potential-energy surface might well lead to good agreement without any tunneling correction.

One important implication of the calculations by Sharp and Johnston is that the predictions based on tunneling through a one-dimensional barrier appear to be too small; for a satisfactory treatment it seems to be necessary to take into account the variation of potential energy with coordinates other than the reaction coordinate.   A suggestion that this is the case is provided by the fact that the difference in tunneling between

$$CH_3 + DH \rightarrow CH_3D + H$$
$$CH_3 + D_2 \rightarrow CH_3D + D$$

is just as great as between

$$CH_3 + HD \rightarrow CH_3H + D$$
$$CH_3 + D_2 \rightarrow CH_3D + D$$

[1] H. S. Johnston, *Advan. Chem. Phys.*, **3**:131 (1960); H. S. Johnston and D. Rapp, *J. Am. Chem. Soc.*, **83**:1 (1961); T. E. Sharp and H. S. Johnston, *J. Chem. Phys.*, **37**:1541 (1962).

[2] *Ibid.*; see also M. J. Stern and M. Wolfsberg, *J. Chem. Phys.*, **45**:4105 (1966).

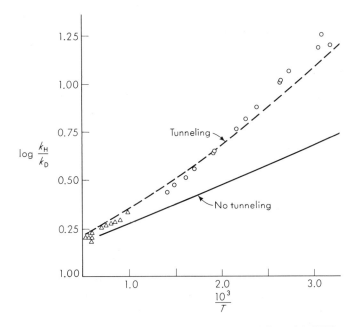

Fig. 35 Arrhenius plot of k_H/k_D for the reaction of CF_3 with CHD_3 (circles) and of CF_3 with CH_2D_2 (triangles). The lines are calculated on the basis of a Sato surface, with and without tunneling; the calculations were for the complete nine-atom system.

This result implies that one cannot simply confine attention to the transfer of the H or D atom, i.e., to the reaction coordinate; the other normal-mode vibrations evidently play an important role. Again, it is to be emphasized that alternative interpretations, not involving tunneling, are possible.

Conclusions   It will be evident from the above account that whereas the activated-complex theory provides a convenient framework for a consideration of kinetic isotope effects and of tunneling, the situation still remains obscure.   The difficulties are twofold and closely related to one another.   One is the problem of obtaining a reliable potential-energy surface; the other, the problem of calculating the tunneling through a multi-dimesnional barrier.   The situation at the present time is that activated-complex theory appears to provide a suitable basis for dealing with kinetic isotope and related effects, but that, until improvements have been made in the calculation of potential-energy surfaces and in the theory of tunneling, it will not be possible to reach firm conclusions.   The evidence tends to suggest that tunneling plays a role in the transfer of H atoms, but is still not completely unequivocal.

TRIMOLECULAR REACTIONS

Only a few elementary trimolecular reactions in the gas phase have been investigated. A very brief account of them will be given here, since details are available elsewhere.[1]

Examples of reactions of this type are

$$2NO + Cl_2 \rightarrow 2NOCl$$

and

$$2NO + O_2 \rightarrow 2NO_2$$

The second reaction is unusual in that the rate decreases with increasing temperature. Attempts by Bodenstein and others[2] to interpret the kinetics of these reactions in terms of a hard-sphere theory of collisions were quite unsuccessful and led to calculated rates which were much too high. This can be understood because there is a considerable loss of entropy when three molecules come together; this loss is not taken into account in a theory of collisions in which the structure of the reactant molecules is neglected.

It is possible that these trimolecular reactions proceed in stages, involving preequilibria such as

$$NO + Cl_2 \rightleftharpoons NOCl_2$$

the species $NOCl_2$ subsequently reacting with a second molecule of nitric oxide,

$$NOCl_2 + NO \rightarrow 2NOCl$$

If this is so, the reactions are not strictly speaking elementary. However, in activated-complex theory the existence of such preequilibria has no effect on the overall kinetics. The concentration of activated complexes can as usual be expressed in terms of the concentration of reactants by using equilibrium theory, and this concentration is not altered if preequilibria of this type are established.

For a reaction

$$A + B + C \rightarrow X^{\ddagger} \rightarrow products$$

[1] See, for example K. J. Laidler, "Chemical Kinetics," pp. 137–143, McGraw-Hill Book Company, New York, 1965.
[2] M. Bodenstein, *Z. physik. Chem.*, **106**:118 (1922); cf. M. Trautz, *Z. Elektrochem.*, **22**:104 (1916); K. F. Herzfeld, *Z. Physik*, **8**:132 (1921).

the rate constant is given by

$$k = l^{\ddagger} \frac{\mathbf{k}T}{h} \frac{Q_{\ddagger}^0}{Q_A{}^0 Q_B{}^0 Q_C{}^0} e^{-E_0/RT} \tag{15}$$

where $l^{\ddagger}$ is the statistical factor and the Q^0's are the partition functions without their symmetry numbers. Since the partition functions are strongly temperature-dependent and there are three factors in the denominator, this equation leads to a fairly strong inverse temperature dependence of the preexponential factor. Thus for reactions of the type

$$2NO + X_2 \rightarrow 2NOX$$

the resulting rate constant has the form

$$k = k' T^{-7/2} e^{-E_0/RT} \tag{16}$$

If, therefore, the energy E_0 is small or zero, the rate constant will decrease with temperature, as was actually found in the reaction involving oxygen.

Detailed calculations of frequency factors on the basis of Eq. (15) have led to values in very satisfactory agreement with experiment.[1] For this small group of reactions activated-complex theory has so far provided the only reliable treatment.

[1] H. Gershinowitz and H. Eyring, *J. Am. Chem. Soc.*, **57**:985 (1935).

6

Unimolecular Reactions

A unimolecular reaction is one in which the activated complex is produced from a single reactant molecule. The formation of this activated complex involves intermolecular energy transfer, and intramolecular energy transfer after the collisions have taken place. At sufficiently high pressures a unimolecular reaction is of the first order, and under these circumstances the activated complexes are at equilibrium (in the same sense as with bimolecular reactions); activated-complex theory can then be applied to the reaction. At lower pressures, however, the kinetics become second order, and equilibrium no longer exists. The rate of reaction is controlled by the rate of energy transfer, both during the course of the energizing collisions and during the subsequent vibration of the energized molecule. The situation is then much more difficult to treat theoretically; a number of alternative interpretations have been proposed, and many important aspects of unimolecular reactions are still not properly understood.

Since the theories of unimolecular reactions are closely related to the treatment of molecular vibrations, the first part of this chapter deals in some detail with this problem. In particular, the equations of motion are worked out explicitly for the vibrations of a linear triatomic molecule, since a consideration of this problem brings out clearly the concept of normal modes and the question of energy flow between the modes.

THE VIBRATIONS OF MOLECULES

Consider first the simple problem of a particle of mass m attached by a weightless spring to a body of infinite mass; the system is shown in Fig. 36. Simple harmonic motion occurs if the restoring force is proportional to the displacement x along the X axis; the equation of motion is

$$m\ddot{x} = -kx \tag{1}$$

The general solution of this type of equation is

$$x = A \cos 2\pi \nu t \tag{2}$$

Double differentiation leads to

$$\ddot{x} = -4\pi^2 \nu^2 A \cos 2\pi \nu t \tag{3}$$
$$= -4\pi^2 \nu^2 x \tag{4}$$

Comparison with Eq. (1) shows that

$$k = 4\pi^2 \nu^2 m \tag{5}$$

so that the frequency of the motion is

$$\nu = \frac{1}{2\pi} \sqrt{\frac{k}{m}} \tag{6}$$

The diatomic molecule   The case of two masses m_1 and m_2 connected by a spring is represented in Fig. 37.   The center of mass of the body is taken as the origin.   If the distance between the particles when the system is at rest is a, the center of mass is at distances $m_2 a/(m_1 + m_2)$ and $m_1 a/(m_1 + m_2)$ respectively from the two masses m_1 and m_2.   Only displacements of the masses along the axis of the molecule need be considered; other displacements correspond to rotation.

Suppose that at a particular time during vibration the mass m_1 is

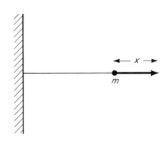

Fig. 36 Simple harmonic motion; a body of mass m connected by a spring of force constant k to a body of infinite mass.

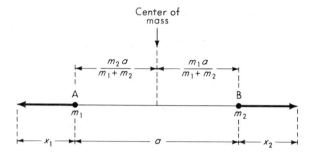

Fig. 37   Vibration of a diatomic molecule.

displaced to the left by a distance x_1 and m_2 is displaced by x_2 to the right.   If moments are taken about the center of mass,

$$m_1 \left(\frac{m_2}{m_1 + m_2} a + x_1 \right) = m_2 \left(\frac{m_1}{m_1 + m_2} a + x_2 \right) \tag{7}$$

whence

$$m_1 x_1 = m_2 x_2 \tag{8}$$

If x is the extension of the spring,

$$x_1 + x_2 = x \tag{9}$$

so that

$$x_1 = \frac{m_2}{m_1 + m_2} x \tag{10}$$

and

$$x_2 = \frac{m_1}{m_1 + m_2} x \tag{11}$$

For harmonic motion the restoring force on each mass is proportional to the extension x; if the proportionality constant is k, the equation of motion for particle A is

$$m_1 \ddot{x}_1 = -kx \tag{12}$$

Double differentiation of (10) gives

$$\ddot{x}_1 = \frac{m_2}{m_1 + m_2} \ddot{x} \tag{13}$$

so that

$$\frac{m_1 m_2}{m_1 + m_2} \ddot{x} = -kx \tag{14}$$

If a reduced mass m is defined by

$$m = \frac{m_1 m_2}{m_1 + m_2} \tag{15}$$

Eq. (13) reduces to

$$m\ddot{x} = -kx \tag{16}$$

which is identical with Eq. (1). Equations (2) to (6) therefore apply equally well to this system, the only modification being that the mass that is involved is the reduced mass defined by Eq. (15).

The linear triatomic molecule   More complicated molecules can be treated by an extension of the above methods.   The equations, however, become very cumbersome for molecules containing more than a few atoms, and systems of simultaneous differential equations have to be solved.   A number of valuable procedures for simplifying the calculations have been worked out and are described elsewhere.[1]   It will suffice for present purposes to consider only the linear triatomic molecule.   The treatment of this will bring out the concept of *normal modes of vibration*, which play a very important role with unimolecular reactions.

The linear triatomic system is shown schematically in Fig. 38.   Consideration will only be given to displacements along the axis of the molecule, which means that bending vibrations (and rotations) are ignored. The center of mass is again taken as the origin in order to eliminate translations of the whole molecule.   If moments are taken about the center of mass when the molecule is at rest,

$$m_1(a_{12} - y) = m_2 y + m_3(a_{23} + y) \tag{17}$$

[1] See, for example, E. B. Wilson, J. C. Decius, and P. C. Cross, "Molecular Vibrations," McGraw-Hill Book Company, New York, 1955.

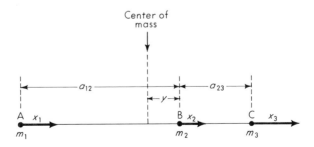

Fig. 38   Stretching vibrations of a linear triatomic molecule.

If the three masses are displaced to the right by the distances x_1, x_2, and x_3,

$$m_1(a_{12} - y - x_1) = m_2(y + x_2) + m_3(a_{23} + y + x_3) \tag{17a}$$

whence

$$m_1 x_1 + m_2 x_2 + m_3 x_3 = 0 \tag{18}$$

The displacements cause the distance between particles A and B to increase by $x_2 - x_1$; the restoring force is therefore $k_{12}(x_2 - x_1)$, where k_{12} is the force constant for the bond. The equation of motion for particle A is therefore

$$k_{12}(x_2 - x_1) = -m_1 \ddot{x}_1 \tag{19}$$

Similarly for particle C

$$k_{23}(x_3 - x_2) = -m_3 \ddot{x}_3 \tag{20}$$

The equation for particle B need not be employed since it is simply a linear combination of Eqs. (18), (19), and (20).

Elimination of x_2 in Eqs. (19) and (20) by using Eq. (18) leads to

$$m_1 m_2 \ddot{x}_1 + k_{12}[(m_1 + m_2)x_1 + m_3 x_3] = 0 \tag{21}$$

and

$$m_2 m_3 \ddot{x}_3 + k_{23}[m_1 x_1 + (m_2 + m_3)x_3] = 0 \tag{22}$$

To solve these equations, one looks for solutions of the form

$$x_1 = A_1 \cos 2\pi \nu t \tag{23}$$
$$x_3 = A_3 \cos 2\pi \nu t \tag{24}$$

the same frequency ν being taken for both displacements. Double differentiation gives

$$\ddot{x}_1 = -4\pi^2 \nu^2 x_1 = -\lambda x_1 \tag{25}$$
$$x_3 = -4\pi \nu^2 x_3 = -\lambda x_3 \tag{26}$$

where λ has been written for $4\pi^2 \nu^2$. Insertion of these expressions into (21) and (22) leads to two simultaneous equations in x_1 and x_3:

$$[-\lambda m_1 m_2 + k_{12}(m_1 + m_2)]x_1 + k_{12} m_3 x_3 = 0 \tag{27}$$
$$m_1 k_{23} x_1 + [-\lambda m_2 m_3 + k_{23}(m_1 + m_3)]x_3 = 0 \tag{28}$$

Equations of this type are only consistent with one another provided that

$$\begin{vmatrix} -\lambda m_1 m_2 + k_{12}(m_1 + m_2) & k_{12} m_3 \\ m_1 k_{23} & -\lambda m_2 m_3 + (m_2 + m_3)k_{23} \end{vmatrix} = 0 \tag{29}$$

This equation is a quadratic equation in λ:

$$\lambda^2 - \left(\frac{m_1 + m_2}{m_2 m_3} k_{12} + \frac{m_2 + m_3}{m_2 m_3} k_{23} \right) \lambda$$

$$+ \frac{k_{12} k_{23}(m_1 + m_2 + m_3)}{m_1 m_2 m_3} = 0 \quad (30)$$

Its solution gives two values of λ and therefore two frequencies ν_1 and ν_2. To simplify the algebra, Eq. (30) may be written as

$$\lambda^2 - b\lambda + c = 0 \quad (31)$$

and the roots are

$$\lambda_1 = \frac{b - \sqrt{b^2 - 4c}}{2} \quad (32)$$

and

$$\lambda_2 = \frac{b + \sqrt{b^2 - 4c}}{2} \quad (33)$$

Since $\lambda_2 > \lambda_1$, the frequency ν_2 is the larger of the two frequencies. If expression (33) for λ is inserted into either Eq. (27) or (28), there results an equation from which it is possible to evaluate the ratio x_1/x_3; when this is done, x_1/x_3 is found to have a positive value. If, on the other hand, the lower root λ_1 is inserted into (27) or (28), the ratio x_1/x_3 is found to be negative.

The significance of this result is that the higher frequency ν_2 (corresponding to λ_2) is the frequency of an asymmetric vibration in which the masses m_1 and m_3 are simultaneously displaced in the same direction; such a vibration is represented in Fig. 39a. The lower frequency corresponds to a symmetric vibration (Fig. 39b) in which an increase in x_1 occurs when there is a decrease in x_3, and vice versa.

It is to be noted that the ratio x_1/x_3 has a fixed value corresponding to the frequency ν_1 and another fixed value corresponding to ν_2. By Eqs.

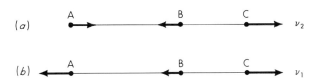

Fig. 39 The two normal modes for the stretching vibrations of a linear triatomic molecule: (a) the asymmetric vibration corresponding to the higher frequency ν_2; (b) the symmetric (breathing) vibration corresponding to the lower frequency ν_1.

(23) and (24) the ratio of amplitudes A_1/A_3 is therefore fixed for each of the frequencies. We may write this amplitude ratio as A_{11}/A_{31} when the symmetric vibration of frequency ν_1 is referred to and as A_{12}/A_{32} for the asymmetric vibration of frequency ν_2. It is also readily shown that the same applies to x_2 and the corresponding amplitude A_2. In other words, $A_{11}/A_{21}/A_{31}$ is independent of the amount of energy in the symmetric vibration, and $A_{12}/A_{22}/A_{32}$ has a fixed value independent of the energy in the asymmetric vibration.

Normal modes of vibration   This example has led in a simple way to the concept of normal modes of vibration. It has been seen that the motions corresponding to the stretchings and contractions of bonds can be reduced to two simple types of harmonic motion, the actual linear vibration of the molecule being a superposition of these two modes. In each of the modes all three atoms are either completely in phase or completely out of phase with one another and move with the same frequency; the motion of each atom, with reference to the center of mass, corresponds to simple harmonic motion. The amplitudes of the vibrations in the two normal modes are in general different from one another; when a molecule becomes excited by collisions, one normal mode may receive more energy than another. The way in which the motion of the atoms is a superposition of the motions corresponding to the two modes is shown in Fig. 40.

A useful way of considering the problem of normal modes for the linear triatomic molecule is with a potential-energy surface, as shown in Fig. 41. The molecule can be regarded as existing in a symmetrical potential-energy basin. The symmetric vibration of frequency ν_1 corresponds to a motion along the line bisecting the X and Y axes, while a motion at right angles to this line corresponds to the asymmetric vibration.

Suppose that the molecule is distorted to a configuration corresponding to point a in the diagram and that the motion is strictly harmonic. The molecule will execute a motion which is a superposition of the two normal modes and in which the configuration at all times remains within the shaded rectangle shown in the diagram. It is important to note that configurations such as those represented by the points b and c, which lie outside the rectangle, are inaccessible to the molecule, even though these points correspond to a lower total energy than that possessed by the molecule.

This latter point is of considerable importance for the theories of unimolecular reactions, as will be discussed in more detail later. N. B. Slater has for the most part made the assumption that the motions are strictly harmonic, which means that once the molecule has started to vibrate, the amounts of energy (and the amplitudes) corresponding to the various normal modes remain the same; this leads to the result that

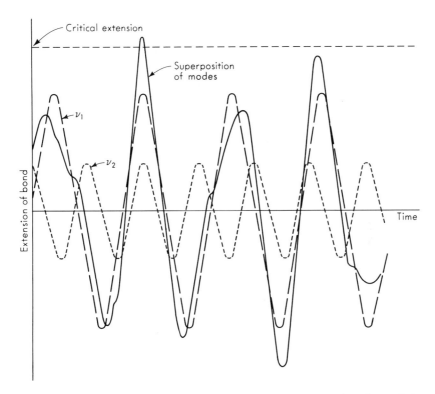

Fig. 40   The extension of a bond due to its motion in two normal modes, and the resultant extension.

points outside the rectangle in Fig. 41 are inaccessible even though the energy requirement is satisfied.   In the theories of Hinshelwood, Kassel, Rice and Ramsperger, and Marcus, on the other hand, there is no such restriction; the normal modes are to be regarded as loosely coupled (which means that they are not true normal modes, the motions being somewhat anharmonic), and in this case energy can flow from one mode to another. The result is that if the molecule starts to vibrate at point a in Fig. 41, all points inside the elliptical contour line through a are accessible; in this case, therefore, the molecule could reach the configurations represented by b and c.

For more complex molecules the situation is, of course, more complicated, but the general principles remain the same.   For a nonlinear molecule containing N atoms there are $3N - 6$ normal modes of vibration; for a linear molecule there are $3N - 5$ (and one less degree of rotational freedom).   In each of the normal modes every atom moves with the same frequency and passes through its equilibrium position (with

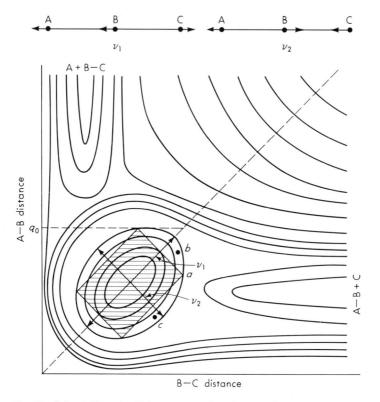

Fig. 41 Schematic potential-energy surface for the linear symmetric molecule A—B—C (A and C identical), showing the axial normal modes of vibration.

respect to the center of mass) at the same time. The actual vibrational motion is again a superposition of all the normal modes. There again remains the important question, to be discussed later in this chapter, of whether or not energy flows between the normal modes; for complicated molecules this makes a considerable difference as far as decomposition is concerned.

LINDEMANN'S THEORY

All modern theories of unimolecular reactions are based on an idea suggested by Lindemann,[1] whose argument can be expressed in the following

[1] F. A. Lindemann, *Trans. Faraday Soc.*, **17**:598 (1922); see M. H. Back and K. J. Laidler (eds.), "Selected Readings in Chemical Kinetics," p. 93, Pergamon Press, 1967.

way.   Occasionally when a collision occurs between two molecules, one of them may acquire a critical amount of energy, sufficient to enable it to become a product molecule without the necessity of acquiring any additional energy; it is then said to be *energized*.   If the conversion of energized molecules into products is slow compared with the rate with which they are deenergized by collision, an equilibrium concentration of them will rapidly be established and their concentration will be proportional to that of the normal molecules.   The rate of reaction, being proportional to the concentration of energized molecules, will thus be proportional to the concentration of normal molecules; the reaction will therefore be of the first order.   At sufficiently low pressures, however, this situation will no longer exist.   The collisions cannot maintain a supply of energized molecules, and the rate of reaction will depend upon the rate of energization and therefore be proportional to the square of the concentration of reactant molecules.

Lindemann's scheme can be formulated as follows.   The process of energization by collision may be represented by the equation

1. $A + A \xrightarrow{k_1} A^* + A$

In this equation A represents a normal reactant molecule, and A^* an energized molecule.   The subsequent reaction of the activated molecule may be represented by the equation

2. $A^* \xrightarrow{k_2}$ products

In addition an energized molecule A^* may suffer deenergization by collision with a normal molecule:

1′. $A^* + A \xrightarrow{k_{-1}} A + A$

The rate of energization by reaction 1 is equal to $k_1[A]^2$, and the rate of deenergization by reaction 1′ is $k_{-1}[A^*][A]$, where k_1 and k_{-1} are the rate constants corresponding to the two processes.   The rate of reaction 2 is equal to $k_2[A^*]$.   Application of the steady-state treatment to $[A^*]$ gives the equation

$$\frac{d[A^*]}{dt} = k_1[A]^2 - k_{-1}[A^*][A] - k_2[A^*] = 0 \tag{34}$$

and from this it follows that

$$[A^*] = \frac{k_1[A]^2}{k_{-1}[A] + k_2} \tag{35}$$

The rate of formation of products is therefore

$$v = k_2[A^*] = \frac{k_1 k_2 [A]^2}{k_{-1}[A] + k_2} \tag{36}$$

At sufficiently high pressures the term $k_{-1}[A]$ is much greater than k_2, which can therefore be neglected; under these conditions the rate is given by

$$v = \frac{k_1 k_2}{k_{-1}} [A] = k_\infty [A] \tag{37}$$

In this equation k_∞ is the first-order constant at high pressure and is equal to $k_1 k_2 / k_{-1}$. The kinetics at high pressures are of the first order. At low pressures, on the other hand, $k_{-1}[A]$ will be negligible compared with k_2 and can be neglected; the rate equation then reduces to

$$v = k_1[A]^2 \tag{38}$$

so that the reaction is now of the second order.

This change from first-order to second-order kinetics has been observed experimentally for a number of reactions. Lindemann's theory therefore does give a satisfactory *qualitative* interpretation of unimolecular reactions, but quantitatively it is not completely correct and some important modifications are required. This may best be seen with the following formulations of the basic rate equation.

Suppose that a first-order rate coefficient k^1 is calculated from experimental data; such a rate coefficient is defined by the equation

$$v = k^1[A] \tag{39}$$

Equations (36) and (39) give

$$k^1 = \frac{k_1 k_2 [A]}{k_{-1}[A] + k_2} = \frac{k_\infty}{1 + k_2/k_{-1}[A]} \tag{40}$$

A plot of k^1 against $[A]$ gives rise to a curve of the form shown in Fig. 42. The first-order coefficient k^1 is constant in the higher pressure range, but falls to zero at lower pressures. It is seen from Eq. (40) that k^1 becomes equal to one-half of k_∞ when $k_{-1}[A]$ is equal to k_2. Therefore, if estimates of k_2 and k_{-1} can be made or if they may be obtained from experiment, it should be possible to predict at what pressure the first-order rate coefficient will begin to fall off. It was with such estimates that difficulties with the Lindemann theory were first encountered.

The concentration $[A]_{1/2}$ at which k^1 should become equal to $k_\infty/2$ is defined by the equation

$$k_{-1}[A]_{1/2} = k_2 \tag{41}$$

whence

$$k_1[A]_{1/2} = \frac{k_1 k_2}{k_{-1}} = k_\infty \tag{42}$$

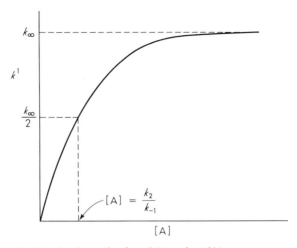

Fig. 42   A schematic plot of k^1 against [A].

or

$$[A]_{1/2} = \frac{k_\infty}{k_1} \qquad (43)$$

The value of k_∞, the first-order rate constant at high pressures, can be obtained from experiment, and according to the simple collision theory k_1 should be equal to $Z_1 e^{-E/RT}$.   However, this procedure always leads to the prediction that the first-order rate constant should fall off at much higher pressures than found experimentally.   Since there can be no doubt about k_∞, which is an experimental quantity, the error must be in the estimation of k_1.   It is therefore necessary to modify the collision theory in such a manner as to give larger values for k_1.

A second difficulty with the Lindemann theory becomes apparent when one plots experimental results in another way.   Equation (40) may be written as

$$\frac{1}{k^1} = \frac{k_{-1}}{k_1 k_2} + \frac{1}{k_1[A]} \qquad (44)$$

and a plot of $1/k^1$ against the reciprocal of the concentration should give a straight line.   However, deviations from linearity have been found of the kind shown schematically in Fig. 43.

HINSHELWOOD'S TREATMENT

The first difficulty with the Lindemann theory, that the first-order rates are maintained down to lower concentrations than the theory appeared

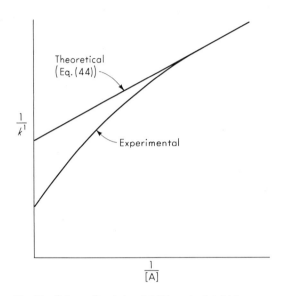

Fig. 43   Schematic plots of $1/k^1$ against $1/[A]$.

to permit, was successfully overcome by Hinshelwood.[1]   The basis of his modification to the Lindemann theory is that the rate constant for the activation process, k_1, may be much greater for a complex molecule than for a simple molecule.   This is so because the energy possessed by a complex molecule may be distributed among a considerable number of degrees of vibrational freedom.   According to statistical mechanics, the fraction f of molecules containing an amount of energy between ε and $\varepsilon + d\varepsilon$ in one degree of freedom is proportional to $e^{-\varepsilon/kT}$;

$$f = ae^{-\varepsilon/kT} \, d\varepsilon \tag{45}$$

where a is the proportionality constant.   Since the total fraction having energy between zero and infinity must be unity,

$$a \int_0^\infty e^{-\varepsilon/kT} \, d\varepsilon = 1 \tag{46}$$

From this, it follows that a is equal to $1/kT$ so that the fraction is given by

$$f = \frac{1}{kT} e^{-\varepsilon/kT} \, d\varepsilon \tag{47}$$

[1] C. N. Hinshelwood, *Proc. Roy. Soc. (London)*, **A113**:230 (1927); see also G. N. Lewis and D. F. Smith, *J. Am. Chem. Soc.*, **47**:1508 (1925); R. H. Fowler and E. K. Rideal, *Proc. Roy. Soc. (London)*, **A113**:570 (1927).

If we consider s degrees of vibrational freedom in a molecule and distribute energy in such a way that the amount in the first degree of freedom is between ε_1 and $\varepsilon_1 + d\varepsilon_1$, that in the second degree of freedom is between ε_2 and $\varepsilon_2 + d\varepsilon_2$, and so on, the fraction is now

$$f = \frac{1}{(\mathbf{k}T)^s} e^{-\varepsilon_1/\mathbf{k}T} e^{-\varepsilon_2/\mathbf{k}T} \cdots e^{-\varepsilon_s/\mathbf{k}T} d\varepsilon_1 d\varepsilon_2 \cdots d\varepsilon_s \tag{48}$$

$$= \frac{1}{(\mathbf{k}T)^s} e^{-\varepsilon/\mathbf{k}T} d\varepsilon_1 d\varepsilon_2 \cdots d\varepsilon_s \tag{49}$$

since the total energy ε is equal to $\varepsilon_1 + \varepsilon_2 + \cdots + \varepsilon_s$. We are here interested, however, in the case where the total energy ε is distributed between the s degrees of freedom in any way, without any restriction as to amounts in the individual degrees of freedom. The integration of

$$d\varepsilon_1 d\varepsilon_2 \cdots d\varepsilon_s$$

subject to the restriction that $\Sigma\varepsilon_i \leq \varepsilon$, proceeds as follows:

$$\int \cdots \int_{\Sigma\varepsilon_1 \leq \varepsilon} d\varepsilon_1 d\varepsilon_2 \cdots d\varepsilon_s = \int_0^\varepsilon d\varepsilon_1 \int_0^{\varepsilon-\varepsilon_1} d\varepsilon_2 \cdots$$
$$\int_0^{\varepsilon-\varepsilon_1 \cdots \varepsilon_{s-1}} d\varepsilon_s \tag{50}$$

$$= \frac{\varepsilon^s}{s!} \tag{51}$$

Differentiation of this expression gives the corresponding integral for the range of energies between ε and $\varepsilon + d\varepsilon$:

$$\iint d\varepsilon_1 d\varepsilon_2 \cdots d\varepsilon_s = d\left(\frac{\varepsilon^s}{s!}\right) = \frac{\varepsilon^{s-1} d\varepsilon}{(s-1)!} \tag{52}$$

The equilibrium proportion F of molecules with total energy between ε and $\varepsilon + d\varepsilon$ and distributed in any way between the s degrees of freedom is the integral of (49) over the range $\varepsilon < \Sigma\varepsilon_i < \varepsilon + d\varepsilon$ and is therefore, by Eq. (52),

$$F = \frac{\varepsilon^{s-1} e^{-\varepsilon/\mathbf{k}T} d\varepsilon}{(s-1)!(\mathbf{k}T)^s} \tag{53}$$

$$= \frac{1}{(s-1)!} \left(\frac{\varepsilon}{\mathbf{k}T}\right)^{s-1} \frac{1}{\mathbf{k}T} e^{-\varepsilon/\mathbf{k}T} d\varepsilon \tag{54}$$

This fraction may be written as dk_1/k_{-1}, dk_1 representing the rate constant for the formation of molecules having energy lying between ε and $\varepsilon + d\varepsilon$; we may therefore write

$$\frac{dk_1}{k_{-1}} = \frac{1}{(s-1)!} \left(\frac{\varepsilon}{\mathbf{k}T}\right)^{s-1} \frac{1}{\mathbf{k}T} e^{-\varepsilon/\mathbf{k}T} d\varepsilon \tag{55}$$

To obtain the value of k_1/k_{-1} this expression must be integrated between ε^* and infinity, where ε^* is the minimum energy that the molecule must have in order for it to decompose into products. We therefore have

$$\frac{k_1}{k_{-1}} = \frac{[\text{A}^*]}{[\text{A}]} = \int_{\varepsilon^*}^{\infty} \frac{1}{(s-1)!} \left(\frac{\varepsilon}{\mathbf{k}T}\right)^{s-1} \frac{1}{\mathbf{k}T} e^{-\varepsilon/\mathbf{k}T} \, d\varepsilon \tag{56}$$

Changing the differential to $de^{-\varepsilon/\mathbf{k}T}$, and integrating by parts gives the solution

$$\frac{k_1}{k_{-1}} = \left[\frac{1}{(s-1)!}\left(\frac{\varepsilon^*}{\mathbf{k}T}\right)^{s-1} + \frac{1}{(s-2)!}\left(\frac{\varepsilon^*}{\mathbf{k}T}\right)^{s-2} \cdots \right] e^{-\varepsilon^*/\mathbf{k}T} \tag{57}$$

For the first term to be much greater than the second, the following condition must be satisfied:

$$\varepsilon^* \gg (s-1)\mathbf{k}T \tag{58}$$

Under these conditions the approximate result of the integration is therefore

$$\frac{k_1}{k_{-1}} = \frac{[\text{A}^*]}{[\text{A}]} = \frac{1}{(s-1)!}\left(\frac{\varepsilon^*}{\mathbf{k}T}\right)^{s-1} e^{-\varepsilon^*/\mathbf{k}T} \tag{59}$$

Since k_{-1} is still considered a collision frequency, k_{-1} is written as Z_{-1} and the expression for k_1 becomes

$$k_1 = Z_{-1} \frac{1}{(s-1)!}\left(\frac{\varepsilon^*}{\mathbf{k}T}\right)^{s-1} e^{-\varepsilon^*/\mathbf{k}T} \tag{60}$$

This expression is to be compared with the expression

$$k_1 = Z_1 e^{-\varepsilon^*/\mathbf{k}T} \tag{61}$$

which is the expression originally employed on the basis of simple collision theory.

The activation energy for k_1 predicted in the Hinshelwood formulation differs from that predicted by the simple collision theory. Equation (60) can be written as

$$k_1 \propto T^{1/2} \left(\frac{1}{T}\right)^{s-1} e^{-\varepsilon^*/\mathbf{k}T} \tag{62}$$

which is to be compared with the corresponding form of Eq. (61)

$$k_{1c} \propto T^{1/2} e^{-\varepsilon^*/\mathbf{k}T} \tag{63}$$

From (62) it is found that

$$\frac{d \ln k_1}{dT} = \frac{(\frac{3}{2} - s)\mathbf{k}T + \varepsilon^*}{\mathbf{k}T^2} \tag{64}$$

Since by the definition of the experimental activation energy (per molecule)

$$\frac{d \ln k_1}{dT} = \frac{\varepsilon_{\exp}}{kT^2} \tag{65}$$

it follows that

$$\varepsilon^* = \varepsilon_{\exp} + (s - \tfrac{3}{2})kT \tag{66}$$

The corresponding treatment applied to Eq. (63) leads to

$$\varepsilon^* = \varepsilon_{\exp} - \tfrac{1}{2}kT \tag{67}$$

It may be shown, by inserting numerical values, that Hinshelwood's theory gives rise to a much higher rate of activation, and therefore to much higher values of k_1/k_{-1}, than does the older theory. If, for example, the experimental activation energy is taken as 40 kcal/mole and s as 12, Eq. (60) gives rise to a value of 9.9×10^{-12} for k_1/k_{-1}. The simple theory, Eq. (61), gives 3.1×10^{-18}. In practice, s is usually found by a method of trial and error, and, with a suitable value for s, it is generally possible to interpret the falling-off in terms of the theory outlined above.

Further developments of the theory are, however, necessary for the following reasons:

1. The number of degrees of freedom, s, cannot be satisfactorily correlated with the total number of degrees of freedom in the molecule; it usually turns out to be smaller. It might be argued that certain degrees of freedom in the molecule do not contribute to s. However a complete theory would be able to predict satisfactorily the number of degrees of freedom to be used. This question will be discussed further with reference to the application of theories to the experimental data.

2. It was mentioned above concerning Eq. (44) that a plot of $1/k^1$ against $1/[A]$ should give a straight line. Hinshelwood's treatment in no way changes this relationship, so that an additional modification to the Lindemann theory is required to take into account the deviations from linearity observed experimentally.

3. Hinshelwood's theory predicts that k_∞ is given by the expression

$$k_\infty = \frac{k_1 k_2}{k_{-1}} = k_2 \frac{1}{(s-1)!} \left(\frac{\varepsilon^*}{kT}\right)^{s-1} e^{-\varepsilon^*/kT} \tag{68}$$

since both k_{-1} and k_2 are considered constant. If this is the case, one would expect a strong temperature dependence of the preexponential factor of the first-order rate constant. On experimental grounds this possibility cannot be denied, since the results do not extend over a sufficiently wide range for an adequate test to be made. However it is to be noted that at high pressures, when equilibrium is established, activated-complex theory should be applicable and the rate constant should be of the form

$$k_\infty = \frac{\mathbf{k}T}{h} \frac{Q^{\ddagger}}{Q_i} e^{-\varepsilon^*/\mathbf{k}T} \tag{69}$$

where Q_i and $Q^{\ddagger}$ are the partition functions for the ground and activated states. The translational and rotational contributions to $Q^{\ddagger}$ and Q_i essentially cancel one another, so that one is left with a ratio of vibrational factors which should not show any strong temperature dependence. A temperature dependence such as is given by Eq. (68) therefore seems to be unlikely.

The treatment of Hinshelwood and the further developments which are now to be discussed are best considered in terms of the following scheme of reactions:

$$A + A \underset{k_{-1}}{\overset{k_1}{\rightleftharpoons}} A^* + A$$

$$A^* \overset{k_2}{\rightarrow} A^{\ddagger}$$

$$A^{\ddagger} \overset{k^{\ddagger}}{\rightarrow} \text{products}$$

In this reaction scheme a distinction has been made between an activated molecule, represented by the symbol $A^{\ddagger}$, and an energized molecule, represented by A^*. By an activated molecule $A^{\ddagger}$ is meant one that is passing directly into the final state. An energized molecule is one that has sufficient energy to become an activated molecule without the acquisition of further energy; it must, however, undergo vibrations before it becomes an activated molecule. The essence of Hinshelwood's modifications to the Lindemann theory is that molecules may become energized much more readily than had been considered possible from simple collision theory, but that a long period of time may elapse before an energized molecule can become an activated molecule. Hinshelwood's treatment predicts an abnormally large value for k_1, and it would postulate a low value for k_2 to compensate for this. The theories that are now to be discussed also postulate a large value for k_1—in fact some of them employ the same expression as in the Hinshelwood treatment—but, instead of regarding k_2 as constant, they consider that k_2 is larger the greater the amount of energy that resides in the energized molecule.

The overcoming of these difficulties has presented a somewhat formidable problem, and a considerable number of attacks have been made.

These are conveniently divided into *statistical* and *dynamical* treatments. The statistical treatments place emphasis on the statistical distribution of energy between the normal modes of the vibrating molecule, the assumption being made that energy flows freely between the modes. The first theories of this kind were presented independently by O. K. Rice and H. C. Ramsperger and by L. S. Kassel; their treatments were very similar to each other and certainly represented an important step in the right direction. Their procedure, now usually referred to as the *RRK treatment*, was improved and developed in 1950 by O. K. Rice and R. A. Marcus and further improved in a series of papers by Marcus and coworkers; these treatments are now usually known collectively as *RRKM treatments*. Important contributions to RRKM theory have also been made by B. S. Rabinovitch. These statistical treatments are discussed in some detail in the present chapter (pages 117 to 129). An important aspect of them is that energy is assumed to flow freely between the normal modes of vibration.

The dynamical treatments of unimolecular reactions start from a completely different point of view in that they treat the detailed dynamical processes of vibration. An early theory of this kind was put forward in 1928 by M. Polanyi and E. P. Wigner, but the first thorough treatment from this point of view was made by N. B. Slater in a series of papers extending from 1939 to the present time. Slater's work has made important contributions to the understanding of molecular vibration and decomposition. The mathematical difficulties in this type of treatment are very great, and as a result most of the work had to be done for strictly harmonic classical motion, which means (see page 107) that energy is not allowed to flow between the modes. For a time it appeared that calculations based on Slater's theory led to a satisfactory interpretation of unimolecular kinetics. Unfortunately, however, it has become apparent that the assumption of no energy flow is an unsatisfactory one and leads to considerable error. More recently Slater and also M. Solc have developed the dynamical treatments in a way that removes the assumption about energy flow; when this is done the treatments give rise to results which more closely resemble the results of RRK theory. These matters are all discussed in more detail later (page 143).

Another type of dynamical treatment of unimolecular reactions was initiated in 1962 by D. L. Bunker. His treatment is related to a specific potential-energy surface for the system. This is a more realistic procedure than that of Slater, which is not explicitly related to a potential-energy surface; however, the calculations are very difficult to carry out and have so far only been done for very simple molecules. Some of the main conclusions from these calculations are briefly summarized in the next chapter.

THE RRK THEORY

The statistical theory of Kassel[1] and of Rice and Ramsperger[2] is based on the assumption that k_2 will be a function of the energy possessed by the energized molecule A*.

According to RRK theory a molecule is assumed to be a system of loosely coupled oscillators; these oscillators are conveniently regarded as equivalent to the normal modes of vibration of the molecule, although this is not an essential requirement; alternatively they may be regarded as individual vibrating bonds. The postulate that the oscillators are loosely coupled is introduced so as to allow a flow of energy between the normal modes, without at the same time destroying the separateness of the normal modes. The oscillators are regarded as all having the same frequency of vibration.

In the RRK theories the rate constant k_2 for the decomposition or isomerization of the active molecule is regarded as increasing with the energy possessed by the molecule in its various degrees of freedom. The larger the energy possessed by the excited molecule, the greater is the chance that it can pass into the bond that is to be broken, and the greater is therefore the rate of the decomposition. The statistical weight of a system of s degrees of vibrational freedom containing j quanta of vibrational energy is equal to the number of ways in which j objects can be divided among s boxes, each of which can contain any number; the number of such ways is

$$w = \frac{(j + s - 1)!}{j!(s - 1)} \tag{70}$$

The statistical weight for states in which the s oscillators have j quanta among them, and one particular one has m quanta, is similarly

$$w' = \frac{(j - m + s - 1)!}{(j - m)!(s - 1)!} \tag{71}$$

The probability that a particular oscillator has m quanta and all s oscillators have j quanta is the ratio of these:

$$\frac{w'}{w} = \frac{(j - m + s - 1)!j!}{(j - m)!(j + s - 1)!} \tag{72}$$

If the Sterling approximation ($n! = n^n/e^n$) is applied to the above expres-

[1] L. S. Kassel, *J. Phys. Chem.*, **32**:225 (1928); "Kinetics of Homogeneous Gas Reactions," chap. 5, Reinhold Publishing Corp., New York, 1932.

[2] O. K. Rice and H. C. Ramsperger, *J. Am. Chem. Soc.*, **49**:1616 (1927); **50**:617 (1928).

sion, the terms in e^n cancel out, and the result is

$$r = \frac{(j - m + s - 1)^{(j-m+s-1)} j^j}{(j - m)^{(j-m)} (j + s - 1)^{(j+s-1)}} \tag{73}$$

Provided that $j - m \gg s - 1$, this reduces to

$$r = \frac{(j - m)^{(j-m+s-1)} j^j}{(j - m)^{(j-m)} j^{(j+s-1)}} \tag{74}$$

$$= \left(\frac{j - m}{j}\right)^{s-1} \tag{75}$$

The total number of quanta j may be taken as proportional to ε, the total energy of the molecule, while m is proportional to ε^*, the minimum energy that a molecule must have for decomposition to take place.[1] The expression given above is therefore equal to

$$r = \left(\frac{\varepsilon - \varepsilon^*}{\varepsilon}\right)^{s-1} \tag{76}$$

The rate with which the required energy ε^* passes into this particular oscillator is proportional to this quantity, so that we may write

$$k_2 = k^{\ddagger} \left(\frac{\varepsilon - \varepsilon^*}{\varepsilon}\right)^{s-1} \tag{77}$$

In this expression $k^{\ddagger}$ is the rate constant corresponding to the free passage of the system over the potential-energy barrier; when ε is sufficiently large, the energized molecule is essentially an activated molecule and therefore can pass immediately into the final state. The variation of k_2 with $\varepsilon/\varepsilon^*$, according to Eq. (77), is shown in Fig. 44.

In order to obtain an expression for k_{∞}, the same expression as Hinshelwood's is used for dk_1/k_{-1}, namely, Eq. (55), and Eq. (77) is used for k_2. The expression for k_{∞} is therefore as follows:

$$k_{\infty} = \frac{k_1 k_2}{k_{-1}} = k^{\ddagger} \int_{\varepsilon^*}^{\infty} \left(\frac{\varepsilon - \varepsilon^*}{\varepsilon}\right)^{s-1} \frac{1}{(s - 1)!} \left(\frac{\varepsilon}{\mathbf{k}T}\right)^{s-1} \frac{1}{\mathbf{k}T} e^{-\varepsilon/\mathbf{k}T} \, d\varepsilon \tag{78}$$

To integrate this expression, the differential may be changed to $de^{-\varepsilon/\mathbf{k}T}$ and the equation rearranged to

$$k_{\infty} = \frac{k^{\ddagger}}{(s - 1)!} \int_{\varepsilon^*}^{\infty} \left(\frac{\varepsilon - \varepsilon^*}{\mathbf{k}T}\right)^{s-1} de^{-\varepsilon/\mathbf{k}T} \tag{79}$$

Substitution of x for $\varepsilon - \varepsilon^*$ in the expression for k_{∞}, with a consequent

[1] This depends upon the assumption that the oscillators all have the same frequency, so that all of the quanta are equal in size.

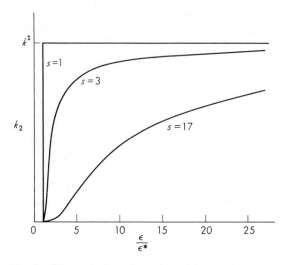

Fig. 44 The variation of k_2 with $\varepsilon/\varepsilon^*$, according to Eq. (77), for different values of s, the number of degrees of freedom.

change in the lower limit, gives

$$k_\infty = \frac{k^{\ddagger}}{(s-1)!} \int_0^\infty \left(\frac{x}{\mathbf{k}T}\right)^{s-1} d\exp\left(-\frac{x+\varepsilon^*}{\mathbf{k}T}\right) \tag{80}$$

$$= \frac{k^{\ddagger}e^{-\varepsilon^*/\mathbf{k}T}}{(s-1)!} \int_0^\infty \left(\frac{x}{\mathbf{k}T}\right)^{s-1} (-de^{-x/\mathbf{k}T}) \tag{81}$$

Integration by parts leads to the result that

$$k_\infty = \frac{k^{\ddagger}e^{-\varepsilon^*/\mathbf{k}T}}{(s-1)!} \left[\left\{-\left(\frac{x}{\mathbf{k}T}\right)^{s-1} - (s-1)\left(\frac{x}{\mathbf{k}T}\right)^{s-2}\right.\right.$$
$$\left.\left.\cdots - (s-1)!\right\} e^{-x/\mathbf{k}T}\right]_0^\infty \tag{82}$$

When the limits are applied to the expression in brackets, the lower limit removes all terms but the last, whereas the upper limit removes all terms without exception; the result is

$$k_\infty = k^{\ddagger}e^{-\varepsilon^*/\mathbf{k}T} \tag{83}$$

This is to be compared with the usual Arrhenius expression for a rate constant, namely,

$$k = Ae^{-\varepsilon^*/\mathbf{k}T} \tag{84}$$

where A is the frequency factor. The Kassel theory does not predict the magnitude of $k^{\ddagger}$, the high-pressure frequency factor, but it is presumably of the order of the vibrational frequency. It is of interest to note that Slater[1] has shown that when Eq. (55) is employed for dk_1/k_1, the only expression that will give rise to Eq. (83) is Eq. (77).

Equation (84) could have been arrived at immediately by using activated-complex theory; according to that theory, $k^{\ddagger}$ would be equal to $\mathbf{k}T/h$, and Eq. (83) would arise provided that the partition functions are the same in the initial and activated states. From the standpoint of activated-complex theory Eq. (83) would therefore predict that frequency factors of first-order reactions should be of the order of $\mathbf{k}T/h$, that is, of the order of 10^{13} at ordinary temperatures. This result is in agreement with experiment for some reactions, but there are a number of exceptions.

The way in which the RRK theories predict the variation of k^1 with the pressure is as follows. Equation (40) can be written in the form

$$k^1 = \frac{k_1 k_2/k_{-1}}{1 + k_2/k_{-1}[A]} \tag{85}$$

and for the rate constant dk^1 associated with molecules having energy lying between ε and $\varepsilon + d\varepsilon$, we may write

$$dk^1 = \frac{(k_2/k_{-1})dk_1}{1 + k_2/k_{-1}[A]} \tag{86}$$

Insertion of expressions (55) and (77) and integration between the limits $\varepsilon*$ and ∞ give rise to

$$k^1 = \int_{\varepsilon*}^{\infty} \frac{\dfrac{1}{(s-1)!}\left(\dfrac{\varepsilon}{\mathbf{k}T}\right)^{s-1} e^{-\varepsilon/\mathbf{k}T} k^{\ddagger} \left(\dfrac{\varepsilon - \varepsilon*}{\varepsilon}\right)^{s-1} \dfrac{d\varepsilon}{\mathbf{k}T}}{1 + \dfrac{k^{\ddagger}}{k_{-1}[A]}\left(\dfrac{\varepsilon - \varepsilon*}{\varepsilon}\right)^{s-1}} \tag{87}$$

This equation may be conveniently reduced by making the following substitutions:

$$x = \frac{\varepsilon - \varepsilon*}{\mathbf{k}T} \qquad b = \frac{\varepsilon*}{\mathbf{k}T} \tag{88}$$

Changing $d\varepsilon/\mathbf{k}T$ to dx and consequently the limits of integration, we obtain

$$k^1 = \int_0^{\infty} \frac{1/[(s-1)!]e^{-(x+b)/\mathbf{k}T}k^{\ddagger}x^{s-1}\,dx}{1 + (k^{\ddagger}/k_{-1}[A])[x/(x+b)]^{s-1}} \tag{89}$$

$$= \frac{k^{\ddagger}e^{-b/\mathbf{k}T}}{(s-1)!}\int_0^{\infty} \frac{x^{s-1}e^{-x}\,dx}{1 + (k^{\ddagger}/k_{-1}[A])[x/(x+b)]^{s-1}} \tag{90}$$

or

$$k^1 = \frac{k^{\ddagger}e^{-\varepsilon*/\mathbf{k}T}}{(s-1)!}\int_0^{\infty} \frac{x^{s-1}e^{-x}\,dx}{1 + (k^{\ddagger}/k_{-1}[A])[x/(x+b)]^{s-1}} \tag{91}$$

---

[1] N. B. Slater, *Proc. Leeds Phil. Lit. Soc.*, **4**:259 (1955).

The integral in the above expressions, for a fixed value of s, corresponds to a particular variation with the concentration [A]. In order to test RRK theory, the procedure is therefore to see, usually by trial and error, what value of s will predict the observed variation of k^1 with the pressure.[1] The application of this theory to a number of reactions has been quite satisfactory. There are still, however, some respects in which the theory is not completely satisfactory. The concept of loosely coupled oscillators, for example, is somewhat vague, and these oscillators are not clearly identifiable with normal modes. Usually Eq. (91) gives satisfactory agreement with experiment if the number of oscillators, s, is taken to be about one-half the total number of normal modes in the molecule, but it is impossible to predict, for a given molecule, the value of s that must be taken. The significance of $k^{\ddagger}$ in the theory is also somewhat unsatisfactory. This quantity is equal to the first-order frequency factor for the reaction, and this for certain molecules is of the order of the average of the vibrational frequencies ($\sim 10^{13}$ sec^{-1}); in such cases it is necessary to suppose that there is a complete redistribution of energy on every vibration. A number of reactions, however, have frequency factors higher by several powers of 10, and RRK theory provides no interpretation of this.

RRKM FORMULATION

Marcus[2] has extended the Rice-Ramsperger-Kassel theory in a direction that brings it into line with activated-complex theory. The essence of his theory, now generally known as the *Rice-Ramsperger-Kassel-Marcus (RRKM)* formulation, is that the individual vibrational frequencies of the energized species and activated complexes are considered explicitly; account is taken of the way the various normal-mode vibrations and rotations contribute to reaction, and allowance is made for the zero-point energies.

As in the previous treatments, the theory relates to the reaction scheme

$$A + A \underset{k_{-1}}{\overset{k_1}{\rightleftharpoons}} A^* + A$$

$$A^* \overset{k_2}{\rightarrow} A^{\ddagger} \rightarrow \text{products}$$

[1] An IBM program for the RRK integral, with tabulated results, has been prepared by E. M. Willbanks, *Los Alamos Lab. Rept. LA*-2178 (1958); it is available from the Office of Technical Services, U.S. Dept. of Commerce, Washington 25, D.C.

[2] R. A. Marcus, *J. Chem. Phys.*, **20**:359 (1952); G. M. Wieder and R. A. Marcus, *J. Chem. Phys.*, **37**:1835 (1962); R. A. Marcus, *J. Chem. Phys.*, **43**:2658 (1965); see also R. A. Marcus and O. K. Rice, *J. Phys. & Colloid Chem.*, **55**:894 (1951).

where A^* represents an energized species, and $A^{\ddagger}$ an activated complex. Energy is allowed to flow freely between normal modes. The rate coefficients k_1 and k_2 are considered functions of the energy of the energized molecules. The first-order rate coefficient that applies to a small energy range is given by

$$dk^1 = \frac{k_2[A]\,dk_1}{k_{-1}[A] + k_2} \tag{92}$$

The true k^1 is then obtained by integration over all possible energies, so that

$$k^1 = \int \frac{k_2[A]\,dk_1}{k_{-1}[A] + k_2} \tag{93}$$

or

$$k^1 = \int \frac{k_2\,dk_1/k_{-1}}{1 + k_2/k_{-1}[A]} \tag{94}$$

In Eq. (94), dk_1/k_{-1} is the fraction of energized molecules having energy in a given energy range. Evaluation of the integrand in Eq. (94) involves a calculation of the number of ways that energy can be distributed among the various degrees of freedom of the energized molecule and the activated complex. The vibrational zero-point energies of A^* and $A^{\ddagger}$ obviously cannot be distributed in this way. Moreover, when the activated complex is formed some energy has gone into the breaking of bonds, and it also is not available for distribution.

Figure 45 is an energy diagram for the reactant species and the resulting activated complex. The total energy contained in the energized molecule, relative to the zero-point level, may be classified as either *active* or *adiabatic*. The *adiabatic* energy is related to those degrees of freedom which remain in the same quantum state during the course of reaction; it contributes no energy to the breaking of bonds. The energy in most of the rotational modes, ε_J and $\varepsilon_J{}^{\ddagger}$, is classed as adiabatic. This is because of the need for conservation of angular momentum; the overall rotational energy in the energized molecule remains as rotational energy in the activated complex, so that it is not available for reaction. In the figure, ε_J is shown as not quite equal to $\varepsilon_J{}^{\ddagger}$; the reason for this is that the structure of the activated complex may be different from that of the energized species, so that the moments of inertia and hence the rotational energies may be different. In the RRKM treatment this change in moments of inertia is treated by introducing, into the rate expression, the ratio $Q_R{}^{\ddagger}/Q_R$ of partition functions corresponding to these adiabatic rotations. Translation of the molecule as a whole also makes no contribution to reaction, again because of the need for conserving momentum; translational energy is therefore adiabatic and is not included.

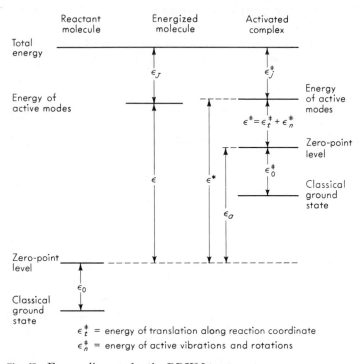

Fig. 45 Energy diagram for the RRKM treatment.

The vibrational modes in the molecule are always treated as *active*; as has been indicated, the evidence is that there is a fairly free flow of energy between the normal modes of vibration, so that all of this energy can make a contribution to reaction. The situation for rotations, however, is not as straightforward. There is not a free interchange of energy between ordinary rotations and vibrations, and, as a result, these rotations usually make little contribution to reaction; there is also, as seen, the need for conservation of rotational momentum. However, certain rotations, in particular some free rotations, can be considered to provide energy to the reaction and are consequently classed as *active*. The question of which rotations are active and which are adiabatic depends upon the particular model chosen for the activated state.

It is seen from Fig. 45 that the active energy ε in the energized molecule is approximately equal to ε^*, the active energy in the activated complex; the procedure involves treating ε^* as the active energy and correcting with the partition-function ratio $Q_R^\ddagger / Q_R$. However, not all of ε^* can pass into vibrational modes of the activated complex because the zero-point level of the latter is higher than that of the reactant molecules.

The true active energy in the activated complex is $\varepsilon^{\ddagger}$, and part of this energy, $\varepsilon_t{}^{\ddagger}$, following activated-complex theory, is to be regarded as corresponding to motion along the reaction coordinate.

The ratio dk_1/k_{-1} is equal to the fraction of molecules having energy in the range of ε^* to $\varepsilon^* + d\varepsilon^*$, where ε^* must be equal to or greater than ε_a (otherwise the molecule is not energized). If $N^*(\varepsilon^*)$ is the number of energy states per unit energy (i.e., the energy density[1]) of the active degrees of freedom, the ratio dk_1/k_{-1} is given by

$$\frac{dk_1}{k_{-1}} = \frac{N^*(\varepsilon^*)e^{-\varepsilon^*/kT}\,d\varepsilon^*}{\int_0^{\infty} N^*(\varepsilon^*)e^{-\varepsilon^*/kT}\,d\varepsilon^*} \tag{95}$$

The denominator of this expression is the partition function corresponding to the active energy contributions; this partition function may be written as Q_a so that

$$\frac{dk_1}{k_{-1}} = \frac{N^*(\varepsilon^*)e^{-\varepsilon^*/kT}\,d\varepsilon^*}{Q_a} \tag{96}$$

Consider now the relationship between the concentrations of activated complexes and energized molecules. The active energy ε^* in the activated complexes is the sum of $\varepsilon_t{}^{\ddagger}$, for its motion along the reaction coordinate, and $\varepsilon_n{}^{\ddagger}$, which relates to the vibrational modes and to any active rotations. The density of quantum states at energy $\varepsilon_t{}^{\ddagger}$ for this translational motion may be written as $N_1(\varepsilon_t{}^{\ddagger})$, and the corresponding density for the vibrational modes, as $N_2(\varepsilon^{\ddagger} - \varepsilon_t{}^{\ddagger})$. Let $[dA^{\ddagger}]$ be the concentration of activated complexes having translational energy between $\varepsilon_t{}^{\ddagger}$ and $\varepsilon_t{}^{\ddagger} + d\varepsilon_t{}^{\ddagger}$ and vibrational energy between $\varepsilon^{\ddagger} - \varepsilon_t{}^{\ddagger}$ and $\varepsilon^{\ddagger} - \varepsilon_t{}^{\ddagger} - d\varepsilon_t{}^{\ddagger}$. If there were a true equilibrium between energized and activated complexes, the ratio of the concentrations $[dA^{\ddagger}]$ and $[A^{\ddagger}]$ would be[2]

$$\frac{[dA^{\ddagger}]}{[A^*]} = \frac{N_1(\varepsilon_t{}^{\ddagger})N_2(\varepsilon^{\ddagger} - \varepsilon_t{}^{\ddagger})\,d\varepsilon_t{}^{\ddagger}}{N^*(\varepsilon^*)} \tag{97}$$

The frequency of passage of activated complexes over the barrier is equal to the rate of passage $\dot{x}$ divided by the arbitrary length δ of the barrier. The rate $\dot{x}$ is related to the energy $\varepsilon_t{}^{\ddagger}$ by

$$\varepsilon_t{}^{\ddagger} = \tfrac{1}{2}m^{\ddagger}\dot{x}^2 \tag{98}$$

[1] If ε_n is the energy level corresponding to a quantum number n, the energy density is $dn/d\varepsilon_n$.

[2] The physical significance of Eq. (97) is that the number of ways that $\varepsilon^{\ddagger}$ can be distributed between the translational and rotational-vibrational levels is the product of the number of ways of putting $\varepsilon_t{}^{\ddagger}$ into the various translational levels and the number of ways of putting $\varepsilon_n{}^{\ddagger}$ ($= \varepsilon^{\ddagger} - \varepsilon_t{}^{\ddagger}$) into the rotational-vibrational levels.

where $m^{\ddagger}$ is the reduced mass; hence

$$\dot{x} = \left(\frac{2\varepsilon_t^{\ddagger}}{m^{\ddagger}}\right)^{1/2} \tag{99}$$

The concentration $[dA^{\ddagger}]$ in Eq. (97) is the concentration at complete equilibrium; we are interested only in those complexes, going in one direction, whose concentration $[dA^{\ddagger}]_l$ is $[dA^{\ddagger}]/2$; thus

$$[dA^{\ddagger}]_l = \frac{[dA^{\ddagger}]}{2} = \frac{[A^*]N_1(\varepsilon_t^{\ddagger})N_2(\varepsilon^{\ddagger} - \varepsilon_t^{\ddagger})\, d\varepsilon_t^{\ddagger}}{2N^*(\varepsilon^*)} \tag{100}$$

Since the frequency of passage over the barrier is $(2\varepsilon_t^{\ddagger}/m^{\ddagger})^{1/2}/\delta$, the rate of passage of complexes from left to right is

$$dv_l = \frac{[A^*]\dot{N}_1(\varepsilon_t^{\ddagger})N_2(\varepsilon^{\ddagger} - \varepsilon_t^{\ddagger})(2\varepsilon_t^{\ddagger}/m^{\ddagger})^{1/2}\, d\varepsilon_t^{\ddagger}}{2N^*(\varepsilon^*)\delta} \tag{101}$$

This rate refers only to the activated complexes in the energy range $d\varepsilon_t^{\ddagger}$; the overall rate v_l is obtained by integrating Eq. (101) with $\varepsilon_t^{\ddagger}$ varying from 0 to $\varepsilon^{\ddagger}$. The result is equal to $k_2[A^*]$, the rate of formation of activated complexes from energized molecules, so that

$$k_2 = \frac{\displaystyle\int_0^{\varepsilon^{\ddagger}} N_1(\varepsilon_t^{\ddagger})N_2(\varepsilon^{\ddagger} - \varepsilon_t^{\ddagger})(2\varepsilon_t^{\ddagger}/m^{\ddagger})^{1/2}\, d\varepsilon_t^{\ddagger}}{2N^*(\varepsilon^*)\delta} \tag{102}$$

Correction by the ratio of partition functions, to allow (as previously explained) for changes in moments of inertia on forming the activated state, then gives

$$k_2 = \frac{Q_R^{\ddagger}\displaystyle\int_0^{\varepsilon^{\ddagger}} N_1(\varepsilon_t^{\ddagger})N_2(\varepsilon^{\ddagger} - \varepsilon_t^{\ddagger})(2\varepsilon_t^{\ddagger}/m^{\ddagger})^{1/2}\, d\varepsilon_t^{\ddagger}}{2Q_R N^*(\varepsilon^*)\delta} \tag{103}$$

The density of translational states for passage through the activated state, $N_1(\varepsilon_t^{\ddagger})$, is given by

$$N_1(\varepsilon_t^{\ddagger}) = \frac{dn}{d\varepsilon_t^{\ddagger}} \tag{104}$$

and the energy levels for a particle in a box of length δ are given by

$$\varepsilon_t^{\ddagger} = \frac{n^2 h^2}{8\delta^2 m^{\ddagger}} \tag{105}$$

Thus

$$n = \frac{2\sqrt{2}\,\delta(m^{\ddagger})^{1/2}}{h}(\varepsilon_t^{\ddagger})^{1/2} \tag{106}$$

so that

$$N_1(\varepsilon_t^{\ddagger}) = \frac{dn}{d\varepsilon_t^{\ddagger}} = \frac{\delta}{h}\left(\frac{2m^{\ddagger}}{\varepsilon_t^{\ddagger}}\right)^{1/2} \tag{107}$$

Introduction of this expression into (103) leads to

$$k_2 = \frac{Q_R^{\ddagger}\int_0^{\varepsilon} N_2(\varepsilon^{\ddagger} - \varepsilon_t^{\ddagger})\, d\varepsilon_t^{\ddagger}}{Q_R N^*(\varepsilon^*)h} \tag{108}$$

Making the change of variables

$$\varepsilon_t^{\ddagger} = \varepsilon^{\ddagger} - \varepsilon_n^{\ddagger} \tag{109}$$

gives

$$k_2 = -\frac{Q_R^{\ddagger}\int_{\varepsilon^{\ddagger}}^0 N_2(\varepsilon_n^{\ddagger})\, d\varepsilon_n^{\ddagger}}{Q_R N^*(\varepsilon^*)h} \tag{110}$$

$$= \frac{Q_R^{\ddagger}\int_0^{\varepsilon} N_2(\varepsilon_n^{\ddagger})\, d\varepsilon_n^{\ddagger}}{Q_R N^*(\varepsilon^*)h} \tag{111}$$

We now have expressions for dk_1/k_{-1} [Eq. (96)] and k_2 [Eq. (111)], and introduction of these into Eq. (93), with integration over all energies from ε_a to infinity, then gives a general expression for the first-order rate coefficient. The result is

$$k^1 = \frac{Q_R^{\ddagger}}{Q_R Q_a h}\int_{\varepsilon_a}^{\infty} \frac{\int_0^{\varepsilon^{\ddagger}} N_2(\varepsilon_n^{\ddagger})\, d\varepsilon_n^{\ddagger}\, e^{-\varepsilon^*/kT}\, d\varepsilon^*}{1 + k_2/k_{-1}[A]} \tag{112}$$

where k_2 is given by Eq. (111). Since $\varepsilon^* = \varepsilon^{\ddagger} + \varepsilon_a$ and $d\varepsilon^* = d\varepsilon^{\ddagger}$, this reduces as follows:

$$k^1 = \frac{Q_R^{\ddagger}e^{-\varepsilon_a/kT}}{Q_R Q_a h}\int_0^{\infty} \frac{\int_0^{\varepsilon^{\ddagger}} N_2(\varepsilon_n^{\ddagger})\, d\varepsilon_n^{\ddagger}\, e^{-\varepsilon^{\ddagger}/kT}\, d\varepsilon^{\ddagger}}{1 + k_2/k_{-1}[A]} \tag{113}$$

At the high-pressure limit $k_2/k_{-1}[A]$ can be neglected, so that

$$k^{\infty} = \frac{Q_R^{\ddagger}e^{-\varepsilon_a/kT}}{Q_R Q_a h}\int_0^{\infty}\int_0^{\varepsilon^{\ddagger}} N_2(\varepsilon_n^{\ddagger})\, d\varepsilon_n^{\ddagger}\, e^{-\varepsilon^{\ddagger}/kT}\, d\varepsilon^{\ddagger} \tag{114}$$

By using the general relationship

$$\int f(x)\, dx = \mathbf{k}T \int f(x)\, d\frac{x}{\mathbf{k}T} \tag{115}$$

it is easily seen that Eq. (114) can be written in the alternative form

$$k^{\infty} = \frac{\mathbf{k}T}{h}\frac{Q_R^{\ddagger}}{Q_R Q_a}\, e^{-\varepsilon_a/kT}\int_0^{\infty}\int_0^{\varepsilon^{\ddagger}} N_2(\varepsilon_n^{\ddagger})\, d\varepsilon_n^{\ddagger}\, e^{-\varepsilon^{\ddagger}/kT}\, d\frac{\varepsilon^{\ddagger}}{\mathbf{k}T} \tag{116}$$

Equation (114) may be integrated by reversing the order of integration so that the limits for $\varepsilon^\ddagger$ and $\varepsilon_n{}^\ddagger$ become $\varepsilon_n{}^\ddagger$ to ∞ and 0 to ∞ respectively; thus

$$k^\infty = \frac{Q_R{}^\ddagger e^{-\varepsilon_a/kT}}{Q_R Q_a h} \int_{\varepsilon_n{}^\ddagger}^{\infty} \int_0^{\infty} N_2(\varepsilon_n{}^\ddagger)\, d\varepsilon_n{}^\ddagger\, e^{-\varepsilon^\ddagger/kT}\, d\varepsilon^\ddagger \qquad (117)$$

The integral $\int_0^{\infty} N_2(\varepsilon_n{}^\ddagger)\, d\varepsilon^\ddagger$ does not involve $\varepsilon^\ddagger$ and may therefore be taken outside the integral sign:

$$k^\infty = \frac{Q_R{}^\ddagger e^{-\varepsilon_a/kT} \int_0^{\infty} N_2(\varepsilon_n{}^\ddagger)\, d\varepsilon_n{}^\ddagger}{Q_R Q_a h} \int_{\varepsilon_n{}^\ddagger}^{\infty} e^{-\varepsilon^\ddagger/kT} d\varepsilon^\ddagger \qquad (118)$$

$$= \frac{kT}{h} \frac{Q_R{}^\ddagger \int_0^{\infty} N_2(\varepsilon_n{}^\ddagger)\, d\varepsilon_n{}^\ddagger\, e^{-\varepsilon_n{}^\ddagger/kT}}{Q_R Q_a} e^{-\varepsilon_a/kT} \qquad (119)$$

The expression $\int_0^{\infty} N_2(\varepsilon_n{}^\ddagger)\, d\varepsilon_n{}^\ddagger\, e^{-\varepsilon_n{}^\ddagger/kT}$ is simply the partition function for those degrees of freedom in $A^\ddagger$ which are not involved in $Q_R{}^\ddagger$ and do not involve the motion along the reaction coordinate. This partition function may be written as $Q_a{}^\ddagger$ so that

$$k^\infty = \frac{kT}{h} \frac{Q_R{}^\ddagger Q_a{}^\ddagger}{Q_R Q_a} e^{-\varepsilon_a/kT} \qquad (120)$$

The product $Q_R{}^\ddagger Q_a{}^\ddagger$ is the conventional complete-partition function $Q_\ddagger$ for the activated complex, motion along the reaction coordinate being neglected; similarly $Q_R Q_a$ is the complete partition Q_i for the reactant molecules; thus

$$k^\infty = \frac{kT}{h} \frac{Q_\ddagger}{Q_i} e^{-\varepsilon_a/kT} \qquad (121)$$

which is the expression obtained from activated-complex theory.

By considering the actual distribution of vibrational-energy levels Marcus[1] has shown that Eq. (113) reduces to

$$k_1 = \frac{kT}{h} \frac{Q_R{}^\ddagger}{Q_R} \frac{e^{-\varepsilon_0/kT}}{(r/2)!} \int_{\varepsilon_v{}^\ddagger \leq \varepsilon^\ddagger}^{\infty} \frac{[(\varepsilon^\ddagger - \varepsilon_v{}^\ddagger)/kT]^{r/2} P(\varepsilon_v{}^\ddagger) e^{-\varepsilon^\ddagger/kT}}{1 + k_2/k_{-1}[A]}\, d\, \frac{\varepsilon^\ddagger}{kT} \qquad (122)$$

where

$$k_2 = \frac{Q_R{}^\ddagger}{h} \frac{1}{(r/1)!} \sum_{\varepsilon_v{}^\ddagger < \varepsilon_v} \frac{[(\varepsilon - \varepsilon_v{}^\ddagger)/kT]^{r/2} P(\varepsilon_v{}^\ddagger)}{N^*(\varepsilon_a + \varepsilon^\ddagger - \varepsilon_0)} \qquad (123)$$

[1] R. A. Marcus, *J. Chem. Phys.*, **43**:2658 (1965); cf. G. M. Wieder and R. A. Marcus, *J. Chem. Phys.* **37**:1835 (1962).

In these equations

> r = number of active rotations of A
> $\varepsilon_v^{\ddagger}$ = energy of the vth vibrational level of $A^{\ddagger}$ minus the zero-point energy of $A^{\ddagger}$
> $P(\varepsilon_v^{\ddagger})$ = degeneracy of the vth vibrational level

Other energies have been defined earlier (see Fig. 45). In certain cases Eqs. (119) and (120) have to be corrected by statistical factors by using the methods discussed on pages 65 to 75.

As will be discussed in a little more detail later, the RRKM theory has been very successful in interpreting the experimental results for unimolecular reactions. For a variety of reactions it has proved possible to formulate a model for the activated complex that will lead to good agreement with the experimental results on the fall-off of the first-order rate coefficients.

SLATER'S THEORY[1]

The theories of unimolecular reactions that have been considered up to now are *statistical* theories. Slater's theory, on the other hand, is a purely *dynamical* one, and takes explicit account of the vibrations of the reacting molecules. In the RRK and RRKM theories it is assumed that energy flows between the normal modes during the course of vibration. Slater, on the other hand, does not permit energy to flow between normal modes; instead, he regards reaction as occurring when a critical coordinate (e.g., a bond length) becomes extended to a specified extent. Such an extension occurs when different normal modes of vibration come suitably into phase. For the most part Slater has treated the problem classically, but has also given a quantum-mechanical formulation.[2]

Slater's classical theory involves a detailed treatment of molecular vibrations, and for it to be applied to the decomposition of an actual molecule, a complete vibrational analysis of the molecule must be made. Unfortunately this frequently presents a difficulty since the necessary spectroscopic data are not always available. The comparison of the various theories with experiment is made in a later section; here it may simply be remarked that Slater's assumption of no energy flow is not realistic, and a modification of the theory is therefore required. Slater's picture

---

[1] For a general account see N. B. Slater, "Theory of Unimolecular Reactions," Cornell University Press, Ithaca, N.Y., 1959; the theory was published in its original form in 1939 [N. B. Slater, *Proc. Cambridge Phil. Soc.*, **35**:56 (1939)] and developed in many subsequent papers.

[2] N. B. Slater, *Proc. Roy. Soc. Edinburgh*, **64**:161 (1955).

of reaction as occurring when a coordinate becomes suitably extended does, however, seem to be a very realistic one, and his treatments have contributed greatly to our understanding of the molecular dynamics of unimolecular reactions.

In view of the difficult mathematical treatment that is required in a presentation of Slater's theory as applied to the general case, it has been thought expedient to discuss its application first to the dissociation of a diatomic molecule and then to that of a linear triatomic molecule.

The decomposition of a diatomic molecule   The potential-energy curve for a diatomic molecule is shown schematically in Fig. 46.   In Slater's treatment the approximation is made of considering the vibration of the molecule to be purely harmonic, which means that the potential-energy curve is considered to be as shown by the dashed line.   Decomposition is supposed to occur when the distance between the atoms has been increased by a certain critical distance x_0 shown in the figure.

The harmonic vibrations of the diatomic molecule were treated earlier in this chapter, and the molecule was seen to have a single mode of vibration of frequency ν given by Eq. (6).   The velocity of the system is obtained by differentiation of Eq. (2),

$$\frac{dx}{dt} = -2\pi\nu A \sin 2\pi\nu t \tag{124}$$

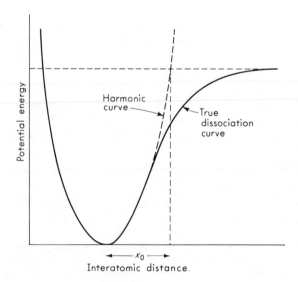

Fig. 46 Potential-energy   curve   for   a   diatomic molecule.

and the maximum velocity, corresponding to zero displacement, is given by

$$\dot{x}_{\text{max}} = \left(\frac{dx}{dt}\right)_{\text{max}} = -2\pi\nu A \tag{125}$$

The maximum kinetic energy of the system is therefore given by

$$\text{KE} = \tfrac{1}{2}m\dot{x}_{\text{max}}^2 = 2m\pi^2\nu^2 A^2 \tag{126}$$

where m is the reduced mass. Since this value corresponds to zero potential energy, it is obviously equal to the total energy of the system, ε. It therefore follows that the half-amplitude A is given by

$$A = \sqrt{\frac{2\varepsilon}{k}} \tag{127}$$

where k is equal to $4\pi^2\nu^2 m$ [cf. Eq. (5)]. Equation (2) may therefore be written as

$$x = \sqrt{\frac{2\varepsilon}{k}} \cos 2\pi\nu t \tag{128}$$

This equation is frequently put into the form

$$x = \alpha \sqrt{\varepsilon} \cos 2\pi\nu t \tag{129}$$

where α is known as the *amplitude factor*. A more general expression is

$$x = \alpha \sqrt{\varepsilon} \cos 2\pi(\nu t + \psi) \tag{130}$$

where ψ takes into account the phase of the vibration.

Slater's method of treating the unimolecular decomposition of a molecule, as applied to this particular case of the diatomic molecule, is that decomposition occurs when the amplitude of vibration reaches a certain critical value x_0, corresponding to energy ε^*, as represented schematically in Fig. 46. The condition that this extension can be attained is given, according to Eq. (130), by the relationship

$$\alpha \sqrt{\varepsilon} \gtrless x_0 \equiv \alpha \sqrt{\varepsilon^*} \tag{131}$$

The critical energy ε^* must be related to this critical extension by

$$\varepsilon^* = \frac{x_0^2}{\alpha^2} \tag{132}$$

The fraction of molecules having the necessary energy ε^*, corresponding to Eq. (132), will be equal to

$$e^{-\varepsilon^*/kT}$$

In this rather trivial case of the unimolecular decomposition of a diatomic molecule it is clear that, if a molecule acquires the energy ε^*, the decomposition will occur in the period of the first vibration, so that the frequency of a decomposition is simply the frequency of the vibration itself. The first-order high-pressure rate constant is therefore given by the expression

$$k = \nu e^{-\varepsilon^*/kT} \tag{133}$$

where ν is the frequency of vibration of the bond that is being ruptured.

The decomposition of a linear triatomic molecule Much more insight into Slater's theory is provided by a consideration of its application to the decomposition of a linear triatomic molecule. The vibrational treatment of this case, for the stretching motions in the molecule, has already been presented (pages 103 to 108). For the application of Slater's treatment it will be assumed that decomposition occurs when q_1, the increase in the distance between atoms A and B, reaches a critical value q_0. From Fig. 38 it is seen that this distance is given by

$$q_1 = x_2 - x_1 \tag{134}$$

The variation of x_1 with time as a result of the first vibrational mode may be represented by the equation [cf. Eq. (23)]

$$x_1 = A_{11} \cos 2\pi\nu_1 t \tag{135}$$

and similar equations for mode 1 apply to x_2 and x_3,

$$x_2 = A_{21} \cos 2\pi\nu_1 t \tag{136}$$
$$x_3 = A_{31} \cos 2\pi\nu_1 t \tag{137}$$

It is not necessary to introduce a phase angle here since the three motions must be in phase. The velocity associated with the motion of atom A in mode 1 is obtained by differentiation of Eq. (135),

$$\dot{x}_1 = -2\pi\nu_1 A_{11} \sin 2\pi\nu_1 t \tag{138}$$

The maximum velocity attained when the atom passes through its equilibrium position is

$$\dot{x}_{1\,max} = -2\pi\nu_1 A_{11} \tag{139}$$

Similar expressions apply to atoms B and C. The energy in the normal mode, equal to the sum of the kinetic energies of the atoms when they pass through their equilibrium positions, is therefore

$$\varepsilon_1 = \tfrac{1}{2}m_1\dot{x}_{1\,max}^2 + \tfrac{1}{2}m_2\dot{x}_{2\,max}^2 + \tfrac{1}{2}m_3\dot{x}_{3\,max}^2 \tag{140}$$
$$= 2\pi^2\nu_1^2(A_{11}^2 m_1 + A_{21}^2 m_2 + A_{31}^2 m_3) \tag{141}$$

We saw earlier that by inserting the value of λ_1 (proportional to the square of the frequency ν_1 for mode 1) into either Eq. (27) or (28) the value of x_1/x_3 can be obtained; the expression is complicated and will not be written down. This ratio x_1/x_3 is actually equal to A_{11}/A_{31}, since x_1 and x_3 represent displacements at any time and A_{11} and A_{31} are the maximum displacements. By writing down the equation in, for example, x_1 and x_2 [analogous to Eq. (27)], it is similarly possible to calculate the ratio x_1/x_2, which is equal to A_{11}/A_{21}. In this way one has therefore obtained the ratios

$$A_{11}/A_{21}/A_{31}$$

Introduction of these ratios into Eq. (141) then allows us to express A_{11}, A_{21}, and A_{31} in terms of the energy ε_1 in the mode; the resulting expressions of course involve the masses and force constants. These expressions may be written in the form

$$A_{11} = \alpha'_{11} \sqrt{\varepsilon_1} \tag{142}$$
$$A_{21} = \alpha'_{21} \sqrt{\varepsilon_1} \tag{143}$$
$$A_{31} = \alpha'_{31} \sqrt{\varepsilon_1} \tag{144}$$

These terms α'_{11}, α'_{21}, and α'_{31} are known as *amplitude factors*, α'_{11}, for example, being related to the contribution made by a given amount of energy in the first normal mode to the displacement of atom A. Similar expressions are found for mode 2:

$$A_{12} = \alpha'_{12} \sqrt{\varepsilon_2} \tag{145}$$
$$A_{22} = \alpha'_{22} \sqrt{\varepsilon_2} \tag{146}$$
$$A_{32} = \alpha'_{32} \sqrt{\varepsilon_2} \tag{147}$$

Since the extension q_1 of the bond between atoms 1 and 2 is equal to $x_2 - x_1$, we may now write, for its variation with time,

$$q_1 = \alpha'_{21} \sqrt{\varepsilon_1} \cos 2\pi(\nu_1 t + \psi_1) + \alpha'_{22} \sqrt{\varepsilon_2} \cos 2\pi(\nu_2 t + \psi_2)$$
$$- \alpha'_{11} \sqrt{\varepsilon_1} \cos 2\pi(\nu_1 t + \psi_1) - \alpha'_{12} \sqrt{\varepsilon_2} \cos 2\pi(\nu_2 t + \psi_2) \tag{148}$$

The phase factors ψ must now be introduced since the two modes do not necessarily start in phase. Equation (148) may be simplified to

$$q_1 = \alpha_{11} \sqrt{\varepsilon_1} \cos 2\pi(\nu_1 t + \psi_1) + \alpha_{12} \sqrt{\varepsilon_2} \cos 2(\nu_2 t + \psi_2) \tag{149}$$

where α_{11} ($= \alpha'_{21} - \alpha'_{11}$) is the amplitude factor which represents the contribution of a given amount of energy in mode 1 to the extension of the critical coordinate in question, and α_{12} is the corresponding contribution from mode 2.

These amplitude factors play a very important part in Slater's theory. It is convenient to define *normalized* amplitude factors μ_1 and μ_2 according to the equations

$$\mu_1 = \frac{\alpha_{11}}{(\alpha_{11}{}^2 + \alpha_{12}{}^2)^{1/2}} \tag{150}$$

and

$$\mu_2 = \frac{\alpha_{12}}{(\alpha_{11}^2 + \alpha_{12}^2)^{1/2}} \tag{151}$$

It is to be noted that

$$\mu_1^2 + \mu_2^2 = 1 \tag{152}$$

These normalized factors still, of course, refer to the extension q_1. Slater defines a mean frequency $\bar{\nu}$ related to this extension by the equation

$$\bar{\nu} = (\mu_1^2 \nu_1^2 + \mu_2^2 \nu_2^2)^{1/2} \tag{153}$$

In view of Eq. (152) this frequency $\bar{\nu}$ must lie between ν_1 and ν_2; it may be regarded as the square root of a weighted mean-square frequency. Insertion of the expressions for μ_1, μ_2, ν_1, and ν_2 leads to the very simple result that

$$\bar{\nu} = \frac{1}{2\pi} \sqrt{\frac{k_{12}}{m}} \tag{154}$$

where

$$m = \frac{m_1 m_2}{m_1 + m_2} \tag{155}$$

In other words, $\bar{\nu}$ is the frequency of the diatomic molecule AB, of atomic masses m_1 and m_2, and of force constant k_{12}; it is therefore the frequency of the system shown in Fig. 38 from which C has been removed. This frequency $\bar{\nu}$ reappears later in the theory [Eq. (170)] as the high-pressure first-order frequency factor.

The process of dissociation of the molecule into atom A and the remaining diatomic species BC is now to be considered. The molecule is assumed to decompose when q_1 attains a specified value q_0. The condition that it can do this is seen from Eq. (149) to be

$$\alpha_{11} \sqrt{\varepsilon_1} + \alpha_{12} \sqrt{\varepsilon_2} \gtrless q_0 \tag{156}$$

The question that we now ask is: What is the minimum value of $\varepsilon_1 + \varepsilon_2$ that just satisfies this condition? The total energy $\varepsilon_1 + \varepsilon_2$ may be written as ε, and the value of ε that just satisfies (156), as ε^*. The minimum value ε^* is obviously one that will cause q_1 to be exactly equal to q_0, so that we may write

$$\alpha_{11} \sqrt{\varepsilon_1} + \alpha_{12} \sqrt{(\varepsilon - \varepsilon_1)} = q_0 \tag{157}$$

The procedure is to minimize ε by using the condition that $d\varepsilon/d\varepsilon_1 = 0$. From Eq. (157) it follows that

$$\varepsilon = \frac{q_0^2 + (\alpha_{11}^2 + \alpha_{12}^2)\varepsilon_1 + 2q_0\alpha_{11}\sqrt{\varepsilon_1}}{\alpha_{12}^2} \tag{158}$$

At the minimum,

$$\frac{d\varepsilon}{d\varepsilon_1} = \frac{\alpha_{11}{}^2 + \alpha_{12}{}^2}{\alpha_{12}{}^2} - \frac{q_0\alpha_{11}}{\alpha_{12}{}^2\varepsilon_1{}^{1/2}} \tag{159}$$

whence

$$\varepsilon_1 = \frac{q_0{}^2\alpha_{11}{}^2}{(\alpha_{11}{}^2 + \alpha_{12}{}^2)^2} \tag{160}$$

$$= \frac{\mu_1{}^2 q_0{}^2}{\alpha_{11}{}^2 + \alpha_{12}{}^2} \tag{161}$$

by using Eq. (150). It is similarly found that, when ε is a minimum,

$$\varepsilon_2 = \frac{\mu_2{}^2 q_0{}^2}{\alpha_{11}{}^2 + \alpha_{12}{}^2} \tag{162}$$

so that the minimum value of ε is given by

$$\varepsilon^* = \frac{\mu_1{}^2 q_0{}^2 + \mu_2{}^2 q_0{}^2}{\alpha_{11}{}^2 + \alpha_{12}{}^2} \tag{163}$$

$$= \frac{q_0{}^2}{\alpha_{11}{}^2 + \alpha_{12}{}^2} \tag{164}$$

since $\mu_1{}^2 + \mu_2{}^2 = 1$. This minimum energy ε^* will later [Eq. (170)] appear as the activation energy for the reaction in the high-pressure region. Equation (164) therefore connects this experimental quantity with the critical extension q_0, in terms of the amplitude factors which can be calculated on the basis of the vibrational analysis of the molecule.

In this connection it is important to distinguish carefully between the definition of energization that is inherent in the RRK theories and that required by Slater. In the RRK theories the condition that

$$\varepsilon \gtrless \varepsilon^* \tag{165}$$

is a sufficient one for energization; the distribution of energy between the normal modes is immaterial because the energy can flow freely between the modes. Slater's theory, on the other hand, imposes in addition to (165) the more stringent requirement that the inequality (156) must be satisfied; by using Eq. (164) this inequality may be expressed as

$$\alpha_{11}\sqrt{\varepsilon_1} + \alpha_{12}\sqrt{\varepsilon_2} \gtrless \sqrt{(\alpha_{11}{}^2 + \alpha_{12}{}^2)\varepsilon^*} \tag{166}$$

As long as (166) is satisfied, (165) must also be satisfied; however, there are distributions of energy that satisfy (165) but not (166).

As a result of this more stringent condition for energization, Slater's theory gives rise to a much lower rate of energization than do the RRK theories. It is useful to compare these two different concepts of energization with reference to the potential-energy surface shown in Fig. 41. We saw that for strict harmonic motion, with no flow of energy, a system starting to vibrate at position a can only execute vibrations represented by motions within the shaded rectangle; thus if the activated state corresponded to either point b or c in the diagram, there could be no reaction, even though the molecule has more total vibrational energy than corresponds to either of the points b and c. Furthermore, if the critical extension of the AB bond is as represented by the horizontal line in the diagram, there can be no reaction. In all cases, on the other hand, the RRK theories, by permitting flow of energy, would allow such molecules to decompose into products.

Slater now treats the general problem of the average frequency (the number of times a second) with which a coordinate q attains the specified value q_0. In a paper[1] published in 1939 he gave an approximate solution to this problem; later, from a general theorem due to Kac,[2] he was able to obtain an exact formula. His result as applied to the present problem of a system with two normal modes is that the average frequency is

$$ L = \left(\frac{\alpha_{11}\sqrt{\varepsilon_1} + \alpha_{12}\sqrt{\varepsilon_2} - q_0}{2\pi}\right)^{1/2} \frac{2}{\sqrt{\pi}} \left(\frac{\alpha_{11}\nu_1^2\sqrt{\varepsilon_1} + \alpha_{12}\nu_2^2\sqrt{\varepsilon_2}}{\alpha_{11}\sqrt{\varepsilon_1}\,\alpha_{12}\sqrt{\varepsilon_2}}\right)^{1/2} \quad (167) $$

This average frequency L is referred to by Slater as the *specific dissociation probability*. This formula of course only applies when the inequality (156) is satisfied; otherwise L is equal to zero.

With this equation it is now possible to obtain expressions for the limiting rate constants at high and low pressures and for the variation with pressure of the first-order rate coefficient. The first-order rate constant for the high-pressure region will be considered first.

The fraction of molecules having energy between ε_1 and $\varepsilon_1 + d\varepsilon_1$ in the first normal mode and between ε_2 and $\varepsilon_2 + d\varepsilon_2$ in the second is given [cf. Eq. (49)] by

$$ f = \frac{e^{-\varepsilon/kT}\,d\varepsilon_1\,d\varepsilon_2}{(kT)^2} \quad (168) $$

This fraction multiplied by the specific dissociation probability and integrated over all energy values is the first-order rate constant; thus

$$ k_\infty = \iint_0^\infty \frac{Le^{-\varepsilon_1/kT}\,d\varepsilon_1\,d\varepsilon_2}{(kT)^2} \quad (169) $$

The lower limit may be set at zero provided that L is taken as zero when the inequality (156) is not satisfied. Insertion of expression (167) for L,

[1] N. B. Slater, *Proc. Cambridge Phil. Soc.*, **35**:56 (1939).
[2] M. Kac, *Amer. J. Math.*, **65**:609 (1943); *Proc. Lond. Math. Soc.*, (2)**50**:390 (1949).

with q_0 written as $\sqrt{(\alpha_{11}^2 + \alpha_{12}^2)\varepsilon^*}$, followed by integration, gives the result that

$$k_\infty = \bar{\nu}e^{-\varepsilon^*/kT} \tag{170}$$

where $\bar{\nu}$ is the average frequency defined by Eq. (153).

An alternative formulation for k_∞ is also of value, especially since it provides a link between Slater's theory and activated-complex theory. It can be shown that the frequency $\bar{\nu}$ is equal to $\nu_1\nu_2/\nu_1^\ddagger$, where ν_1 and ν_2 are the two normal-mode frequencies and $\nu_1^\ddagger$ is the frequency of vibration of the activated complex, in which q_1 is held fixed; $\nu_1^\ddagger$ is in fact given by

$$\nu_1^\ddagger = \frac{1}{2\pi}\sqrt{\frac{k_{23}}{m^\ddagger}} \tag{171}$$

where

$$\frac{1}{m^\ddagger} = \frac{1}{m_1 + m_2} + \frac{1}{m_3} \tag{172}$$

The easiest way to prove this relationship is to show that

$$\bar{\lambda} = \frac{\lambda_1\lambda_2}{\lambda_1^\ddagger} \tag{173}$$

where the λ's are related to the corresponding frequencies by the equation

$$\lambda = 4\pi^2\nu^2 \tag{174}$$

It is seen from the quadratic equation (30) that the product of the roots is

$$\lambda_1\lambda_2 = \frac{k_{12}k_{23}(m_1 + m_2 + m_3)}{m_1m_2m_3} \tag{175}$$

From (171) and (172)

$$\lambda_1^\ddagger = \frac{k_{23}}{m^\ddagger} = \frac{k_{23}(m_1 + m_2 + m_3)}{m_3(m_1 + m_2)} \tag{176}$$

Hence

$$\frac{\lambda_1\lambda_2}{\lambda_1^\ddagger} = \frac{k_{12}(m_1 + m_2)}{m_1m_2} = \bar{\lambda} \tag{177}$$

with the use of (154) and (155). In view of this relationship the rate constant may be written as

$$k_\infty = \frac{\nu_1\nu_2}{\nu_1^\ddagger}e^{-\varepsilon^*/kT} \tag{178}$$

An identical expression is obtained from activated-complex theory if classical vibrational-partition functions $kT/h\nu$ are employed. According to that theory

$$k_\infty = \frac{kT}{h} \frac{Q^\ddagger}{Q_i} e^{-\varepsilon^*/kT} \tag{179}$$

where Q_i and $Q^\ddagger$ are the partition functions for initial and activated states, respectively, and insertion of the classical partition functions gives

$$k_\infty = \frac{kT}{h} \frac{kT/h\nu_1^\ddagger}{(kT/h\nu_1)(kT/h\nu_2)} e^{-\varepsilon^*/kT} \tag{180}$$

$$= \frac{\nu_1\nu_2}{\nu_1^\ddagger} e^{-\varepsilon^*/kT} \tag{181}$$

In order for this formulation to coincide with that of Slater, the same reaction coordinate must of course be chosen; for $\nu_1^\ddagger$ to be equal to (171), the decomposition of the molecule must be regarded as involving only the extension of the A—B bond, so that the relevant partition function relates to the molecule in which the A—B distance is fixed.

The second-order rate constant k_1 is given, according to Slater's theory, by using the relationship

$$\frac{dk_1}{k_{-1}} = \frac{e^{-\varepsilon/kT} d\varepsilon_1 d\varepsilon_2}{(kT)^2} \tag{182}$$

so that, by putting k_{-1} equal to Z_{-1},

$$k_1 = \int_0^\infty dk_1 = \frac{Z_{-1}}{(kT)^2} \int\int_0^\infty e^{-\varepsilon/kT} d\varepsilon_1 d\varepsilon_2 \tag{183}$$

When the integration is performed subject to the inequality (166), the result is

$$k_1 = Z_{-1}\left(\frac{\varepsilon^*}{kT} 4\pi\right)^{1/2} \mu_1\mu_2 e^{-\varepsilon^*/kT} \tag{184}$$

where the μ's are the normalized amplitude factors defined by Eqs. (150) and (151).

The general rate expression is obtained by writing down the appropriate steady-state equations. Consider an energized molecule having energy between ε_1 and $\varepsilon_1 + d\varepsilon_1$ in mode 1 and between ε_2 and $\varepsilon_2 + d\varepsilon_2$ in mode 2. The equilibrium concentration of such molecules is

$$[A^*] = \frac{[A]e^{-\varepsilon/kT} d\varepsilon_1 d\varepsilon_2}{(kT)^2} \tag{185}$$

and the rate of energization of molecules to this level is

$$v_1 = \frac{Z_1[A]^2 e^{-\varepsilon/kT} d\varepsilon_1 d\varepsilon_2}{(kT)^2} \tag{186}$$

The concentration in the steady state will be less than this, owing to the occurrence of reaction; the concentration may be written as

$$[A^*] = g[A] \, d\varepsilon_1 \, d\varepsilon_2 \tag{187}$$

where g is a function to be determined. The rate of deenergization is given by

$$v_{-1} = Z_{-1}[A][A^*] \tag{188}$$
$$= Z_{-1}g[A]^2 \, d\varepsilon_1 \, d\varepsilon_2 \tag{189}$$

The rate of reaction of the energized molecules is equal to their steady concentration multiplied by the specific dissociation probability L:

$$v_2 = Lg[A] \, d\varepsilon_1 \, d\varepsilon_2 \tag{190}$$

The steady-state expression is therefore

$$v_1 = v_{-1} + v_2 \tag{191}$$

or

$$\frac{Z_1[A]^2 e^{-\varepsilon/kT} \, d\varepsilon_1 \, d\varepsilon_2}{(\mathbf{k}T)^2} = Z_{-1}g[A]^2 \, d\varepsilon_1 \, d\varepsilon_2 + Lg[A] \, d\varepsilon_1 \, d\varepsilon_2 \tag{192}$$

The proportionality factor g is found from this to be

$$g = \frac{Z_1[A]e^{-\varepsilon/kT}}{(\mathbf{k}T)^2(L + Z_{-1}[A])} \tag{193}$$

The steady concentration of these particular energized molecules is thus, from Eq. (186),

$$[A^*] = \frac{Z_1[A]^2 e^{-\varepsilon/kT} \, d\varepsilon_1 \, d\varepsilon_2}{(\mathbf{k}T)^2(L + Z_{-1}[A])} \tag{194}$$

and the rate of reaction, equal to $L[A^*]$, is thus

$$v = \frac{LZ_1[A]^2 e^{-\varepsilon/kT} \, d\varepsilon_1 \, d\varepsilon_2}{(\mathbf{k}T)^2(L + Z_{-1}[A])} \tag{195}$$

The above equations apply only to the molecules having energies between ε_1 and $\varepsilon_1 + d\varepsilon_1$ in mode 1 and ε_2 and $\varepsilon_2 + d\varepsilon_2$ in mode 2; the overall rate is obtained by integration, and the first-order rate coefficient, equal to $v/[A]$, is thus

$$k^1 = \int_{\varepsilon_1}^{\infty} \int_{\varepsilon_2}^{\infty} \frac{LZ_1[A]e^{-\varepsilon/kT} \, d\varepsilon_1 \, d\varepsilon_2}{(\mathbf{k}T)^2(L + Z_1[A])} \tag{196}$$

The integration must be carried out subject to the restriction that

$$\varepsilon^* \leq \varepsilon_1 + \varepsilon_2 \leq \varepsilon^* + d\varepsilon^* \tag{197}$$

and the final result is

$$k^1 = \frac{\bar{\nu} e^{-\varepsilon^*/kT}}{\sqrt{\pi}} \int_0^\infty \frac{x^{1/2} e^{-x}\, dx}{1 + x^{1/2}\theta^{-1}} \qquad (198)$$

x is defined, as previously, by

$$x = \frac{\varepsilon - \varepsilon^*}{kT} \qquad (199)$$

and θ is given by

$$\theta = \frac{Z_1[A]2\pi}{\bar{\nu}} \left(\frac{\varepsilon^*}{kT}\right)^{1/2} \mu_1\mu_2 \qquad (200)$$

The general case   The extension of the above treatment to the general case of a polyatomic molecule having n normal modes of vibration[1] is fairly obvious, and only the main results will be stated here.   Suppose that a certain coordinate q is chosen as the reaction coordinate; the extension of this to a critical value q_0 leads to reaction.   It is not necessary for q to be a simple bond length; it can, however, be related linearly to bond lengths (or to normal coordinates), and its variation with time is therefore expressible as [cf. Eq. (149)]

$$q = \Sigma \alpha_r \sqrt{\varepsilon_r} \cos 2\pi(\nu_r t + \psi_r) \qquad (201)$$

the summation sign Σ representing $\displaystyle\sum_{r=1}^{n}$, the summation being taken over all the n normal modes.   The condition that q can attain the value q_0 is [cf. Eq. (156)]

$$\Sigma \alpha_r \sqrt{\varepsilon_r} \gtreqless q_0 \qquad (202)$$

The minimum value of the total energy $\Sigma \varepsilon_r$ that will satisfy this requirement can be shown to be [cf. Eq. (164)]

$$\varepsilon^* = \frac{q_0^2}{\Sigma \alpha_r^2} \qquad (203)$$

Condition (202) is therefore

$$\Sigma \alpha_r \sqrt{\varepsilon_r} \gtreqless \sqrt{\varepsilon^* \Sigma \alpha_r^2} \qquad (204)$$

The specific dissociation probability, which is the number of times a second that the coordinate q reaches the specified value q_0, is then given

---

[1] The symbol n will be used for the number of normal modes in the Slater theory, and s for the number in the RRK formulations; it is convenient to have a separate symbol since the values used in the application of the theories to the data are generally different.

by [cf. Eq. (167)]

$$L = \frac{(\Sigma\alpha_r \sqrt{\varepsilon_r} - q_0)^{(n-1)/2}}{(2\pi)^{(n-1)/2}\Gamma(\frac{1}{2}n + \frac{1}{2})} \frac{\Sigma\alpha_r\nu_r{}^2 \sqrt{\varepsilon_r}}{\alpha_1 \sqrt{\varepsilon_1}\, \alpha_2 \sqrt{\varepsilon_2} \cdots \alpha_n \sqrt{\varepsilon_n}} \tag{205}$$

where the symbol Γ represents the gamma function.[1]

The expression for the high-pressure first-order rate constant is the same as Eq. (170), i.e.,

$$k_\infty = \bar{\nu}e^{-\varepsilon^*/kT} \tag{206}$$

where $\bar{\nu}$, the average frequency, is now defined by [cf. Eq. (153)]

$$\bar{\nu} = (\Sigma\mu_r{}^2\nu_r{}^2)^{1/2} \tag{207}$$

the μ's again being the normalized amplitude factors [cf. Eq. (150)]

$$\mu_1 = \frac{\alpha_1}{(\Sigma\alpha_r{}^2)^{1/2}} \quad \text{etc.} \tag{208}$$

The physical significance of $\bar{\nu}$ is that it is the frequency of the molecule in which all atoms have been removed except those related to the coordinate q; if q is a simple bond, it is the frequency of the diatomic molecule. The frequency ν is also related to other frequencies by the equation

$$\bar{\nu} = \frac{\nu_1\nu_2 \cdots \nu_n}{\nu_2{}^{\ddagger}\nu_3{}^{\ddagger}\nu_4{}^{\ddagger} \cdots \nu_n{}^{\ddagger}} \tag{209}$$

where $\nu_2{}^{\ddagger}$, $\nu_3{}^{\ddagger}$, . . . , $\nu_n{}^{\ddagger}$ are the $n-1$ frequencies which arise if the coordinate q is held fixed. The rate constant may therefore be written as [cf. Eq. (178)]

$$k = \frac{\nu_1\nu_2 \cdots \nu_n}{\nu_2{}^{\ddagger}\nu_3{}^{\ddagger}\nu_4{}^{\ddagger} \cdots \nu_n{}^{\ddagger}} e^{-\varepsilon^*/kT} \tag{210}$$

---

[1] The gamma function is defined by the relationship

$$\Gamma(n+1) = \int_0^\infty e^{-x}x^n \, dx \tag{205a}$$

In the event that n is an integer, the function reduces as follows:

$$\Gamma(n+1) = n! \tag{205b}$$

The quantity $\frac{1}{2}n + \frac{1}{2}$ appearing in Eq. (205) may also be half-integral, and in that event the value of the gamma function may be calculated by making use of the following relationships:

$$\Gamma(n+1) = n\Gamma(n) \tag{205c}$$
$$\Gamma(\tfrac{1}{2}) = \pi \tag{205d}$$

For example, the gamma function of $\frac{7}{2}$ is calculated as follows:

$$\Gamma(\tfrac{7}{2}) = \tfrac{5}{2}\Gamma(\tfrac{5}{2}) = \tfrac{5}{2}\tfrac{3}{2}\Gamma(\tfrac{3}{2}) = \tfrac{5}{2}\tfrac{3}{2}\tfrac{1}{2}\Gamma(\tfrac{1}{2}) = \tfrac{5}{2}\tfrac{3}{2}\tfrac{1}{2}\sqrt{\pi} \tag{205e}$$

and this equation again is equivalent to that given by activated-complex theory if classical partition functions are employed.

The second-order rate constant [cf. Eq. (184)] is given by the expression

$$k_1 = Z_{-1}\left(\frac{\varepsilon^*}{kT}4\pi\right)^{(n-1)/2} \mu_1\mu_2 \cdots \mu_n e^{-\varepsilon^*/kT} \tag{211}$$

The general rate equation is found, by a steady-state argument similar to that of Eqs. (185) to (198), to be

$$k^1 = \frac{\bar{\nu}e^{-\varepsilon^*/kT}}{\Gamma(\frac{1}{2}n + \frac{1}{2})}\int_0^\infty \frac{x^{(n-1)/2}e^{-x}\,dx}{1 + x^{(n-1)/2}\theta^{-1}} \tag{212}$$

where x is $(\varepsilon - \varepsilon^*)/kT$ and θ is defined by

$$\theta = \frac{Z_1[A]}{\bar{\nu}}\left(\frac{\varepsilon^*}{kT}4\pi\right)^{(n-1)/2}\Gamma(\tfrac{1}{2}n + \tfrac{1}{2})\mu_1\mu_2 \cdots \mu_n \tag{213}$$

It follows from (212) and (206) that

$$\frac{k^1}{k_\infty} = \frac{1}{\Gamma(\frac{1}{2}n + \frac{1}{2})}\int_0^\infty \frac{x^{(n-1)/2}e^{-x}\,dx}{1 + x^{(n-1)/2}\theta^{-1}} \tag{214}$$

Slater[1] has tabulated values of this integral for values of n ranging from 3 to 16.

Slater's treatment is of considerable interest, but it has become increasingly clear that his initial assumption of no energy flow is unrealistic. Various attempts have therefore been made to develop a treatment along similar lines to Slater's but without this assumption. Gill and Laidler[2] suggested that this could be done in terms of the following reaction scheme:

[1] N. B. Slater, "Theory of Unimolecular Reactions," p. 169, Cornell University Press, Ithaca, N.Y., 1959.
[2] E. K. Gill and K. J. Laidler, *Proc. Roy. Soc. (London)*, **A250**:121 (1959).

Here a distinction has been made between two types of energized molecules, represented by A′ and A*. The A* molecules are those that are energized in the Slater sense; they not only contain ε^* or more of energy but have it distributed among the normal modes in such a way that, when the vibrations come suitably into phase, there can be a sufficient extension of the critical coordinate. The A′ molecules are those that contain the critical energy ε^*, but do not have it suitably distributed for reaction to occur without flow of energy. The energization rate constant $k_1{}^H$ is much larger than $k_1{}^S$.

If there is no flow of energy between the modes, the rate constants k_s and k_{-s} are equal to zero, and reaction can only occur through direct energization to A*; Slater's treatment is then applicable. If, on the other hand, k_s is not negligible (which seems to be the true situation), A* can be formed not only directly but also from A′. At very low pressures, in fact, most of the A* may be formed from A′, since A′ may be formed from A + A much more rapidly than A*, and at very low pressures practically every A′ formed will eventually become an A*. At intermediate pressures, on the other hand, more of the A* may be produced directly from A + A, and Slater's formulation will then be closer to the truth. From these ideas one would therefore expect three regions of kinetic behavior:

1. A high-pressure, first-order region. Here both A′ and A* will be essentially at equilibrium. In this region the RRK and Slater theories are in agreement. They are still deficient in not allowing for high-frequency factors; these are taken care of in the RRKM and activated-complex formulations by a high entropy of the activated state.
2. An intermediate-pressure region, where there is predominantly a direct energization to form A*. Slater's formulation may be applicable here.
3. A low-pressure region, where A* will be formed predominantly from A′ and the RRK formulation will apply.

Gill and Laidler developed steady-state equations which lead to these results; there is, of course, no sharp division between the regions.

Solc[1] has carried out a mathematical development of Slater's theory on much the same ideas. He obtains a first-order rate coefficient of the

[1] M. Solc, *Mol. Phys.*, **11**:579 (1966); **12**:101 (1967); *Z. Physik. Chem. (Leipzig)*, **234**:185 (1967); *Chem. Phys. Letters*, **1**:160 (1967); cf. N. B. Slater, *Mol. Phys.*, **12**:107 (1967).

form

$$k^1 = \frac{Z[M]}{(kT)^n} \int_0^\infty \cdots \int_0^\infty \frac{L}{L + Z[M] + \omega}$$
$$\frac{\int \cdots \int d\varepsilon_1 \cdots d\varepsilon_s}{\int \cdots \int (L + Z[M])(L + Z[M] + \omega)^{-1} d\varepsilon_1 \cdots d\varepsilon_s}$$
$$e^{-\varepsilon/kT} d\varepsilon_1 \cdots d\varepsilon_n \quad (215)$$

Here Z is the collision number; $[M]$, the reactant concentration; and L, the specific dissociation probability [Eq. (205)]. It is assumed that there are n normal modes altogether and that between s of them the energy can flow with a frequency ω. The results of some calculations for $n = 4$ and $s = 3$, with various values for ω, are shown in Fig. 47. The case of $\omega = 0$ corresponds to the original Slater formulation. When ω is equal to or greater than the vibrational frequencies, the results of the calculations are found to be indistinguishable from those given by the original RRK formulation.

These treatments in which the no-flow restriction is removed have evidently brought about a considerable improvement; we now seem to have the main outlines of a very satisfactory classical dynamical treat-

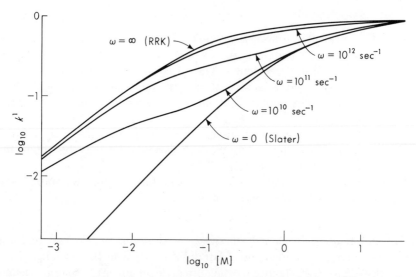

Fig. 47 Calculated dependencies of k^1 on $[M]$, as given by Solc's modification of Slater's theory. The calculations are for four normal modes (n) between three of which (s) energy flow can occur with the frequencies indicated. Except for $\omega = 0$ all of the curves approach the RRK behavior at low pressures. The case of $\omega = 10^{10}$ sec^{-1} is of interest in showing the intermediate-pressure region where the Slater description is close to the truth.

ment of unimolecular reactions. One important improvement that is still called for is the introduction of greater degrees of anharmonicity, so that the theory will deal with cases where there is a considerable loosening of structure in the activated state. A formulation that allows for quantum effects is also called for; Slater[1] has attempted such a treatment, but further work is necessary.

APPLICATION OF THEORIES TO EXPERIMENTAL RESULTS

A vast amount of work has been done on the application of theories of unimolecular reactions to the experimental data. No attempt will be made to give a comprehensive review of this subject.

Energy flow Calculations which had as their object the testing of the hypothesis of no energy flow were carried out by Gill and Laidler[2] for a number of reactions. Emphasis was placed on the low-pressure region, where the Slater theory leads to much lower calculated rates than do formulations in which energy flow is permitted. The calculations showed that for several reactions (e.g., the decomposition of hydrogen peroxide) the Slater treatment was incapable of giving satisfactory agreement with experiment. Similar conclusions have been reached for other reactions (e.g., the cyclobutane decomposition) by Thiele and Wilson[3] and by Chesick.[4]

The conclusion that strictly normal-mode treatments, with no energy flow, are not adequate has received further support from theoretical treatments of vibrational processes, particularly by Thiele and Wilson[5] and by Bunker.[6]

Cyclopropane isomerization A considerable amount of experimental and theoretical work has been done on the thermal structural isomerization

[1] N. B. Slater, *Proc. Roy. Soc. Edinburgh*, **64**:161 (1955).

[2] E. K. Gill and K. J. Laidler, *Proc. Roy. Soc. (London)*, **A250**:121 (1959); **A251**:66 (1959); *Can. J. Chem.*, **36**:1570 (1958); *Trans. Faraday Soc.*, **55**:753 (1959); see also K. J. Laidler and B. W. Wojciechowski, *Chem. Soc. (London) Spec. Publ.*, **16**:37 (1962).

[3] E. Thiele and D. J. Wilson, *Can. J. Chem.*, **37**:1035 (1959); *J. Phys. Chem.*, **64**:473 (1960).

[4] J. P. Chesick, *J. Am. Chem. Soc.*, **82**:2277 (1960).

[5] E. Thiele and D. J. Wilson, *J. Chem. Phys.*, **35**:1256 (1961); E. Thiele, *J. Chem. Phys.*, **38**:1959 (1963).

[6] D. L. Bunker, *J. Chem. Phys.*, **37**:393 (1962); **40**:1946 (1964).

of cyclopropane into propylene,

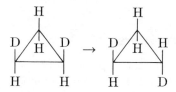

A geometrical isomerization also occurs with compounds such as *cis*-deuteriocyclopropane:

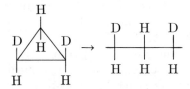

Rabinovitch et al.[1] have found that this geometrical isomerization occurs considerably more rapidly than the structural isomerization into dideuteriopropylene. This result has been regarded as favoring the view that a diradical is formed initially,

and that this can form the cyclic molecule again, with the possibility of geometrical isomerization, or it can form the dideuteriopropylene. However, alternative mechanisms are possible.[2]

Slater[3] applied his theory and the RRK theory to the results of Pritchard, Sowden, and Trotman-Dickenson[4] and found that both theories could be made to fit the fall-off curves; the predictions of the two theories are, in fact, very close together except at pressures of less than 10^{-1} mm. Agreement with the RRK theory was obtained by taking the number of effective degrees of freedom, s, to be 12; the total number in the C_3H_6 molecule is 21. Slater's theory led to good agreement when all 21 modes are taken into consideration and when the reaction coordinate was taken to be the distance between a carbon atom and a hydrogen atom that is attached to one of the other carbon atoms. The choice of a carbon-carbon

[1] B. S. Rabinovitch, E. W. Schlag, and K. B. Wiberg, *J. Chem. Phys.*, **28**:504 (1958).

[2] Mechanistic aspects of the problem are discussed by K. J. Laidler and L. F. Loucks in C. H. Bamford and C. F. Tipper (eds.), "Comprehensive Chemical Kinetics," vol. 5, to be published by Elsevier Publishing Company, Amsterdam.

[3] N. B. Slater, *Proc. Roy. Soc. (London)*, **A218**:224 (1953).

[4] H. O. Pritchard, R. G. Sowden, and A. F. Trotman-Dickenson, *Proc. Roy. Soc. (London)*, **A217**:563 (1953).

distance as reaction coordinate gave much too low a value for the rate of energization and therefore predicted a falling-off of k^1 over much too high a pressure range.

The falling-off in the rate coefficient k^1 for cis-dideuteriocyclopropane has been found[1] to occur in the same pressure range as for cyclopropane. This result has been regarded as an argument against Slater's theory, on the following grounds. In cyclopropane 14 of the 21 vibrational modes are doubly degenerate, and according to Slater's theory each of the 7 pairs can only be counted as one. The physical reason for this is that, if two vibrations are of the same frequency, they can never come into phase if they once start out of phase; they therefore act as one as far as coming into phase with other vibrations is concerned. As a result of this, in cyclopropane only 14 modes (7 nondegenerate modes and 7 of the the 14 doubly degenerate ones) contribute to the lengthening of the critical coordinate. In dideuteriocyclopropane, on the other hand, there are no degenerate vibrations; all 21 should therefore make a contribution, and this seems to lead to the conclusion that the fall-off should occur over a very much lower pressure range than with cyclopropane itself. However, Golike and Schlag[2] have shown this argument to be invalid. They found that when detailed calculations are made for dideuteriocyclopropane some of the amplitude factors (see page 133) have very small values and that the Slater treatment in fact predicts a falling-off over the same pressure range. However, as has been discussed, there is now good evidence against Slater's hypothesis of no energy flow.

Application of the RRKM theory to the cyclopropane isomerization was made by Wieder and Marcus.[3] They obtained excellent agreement with the experimental results with all 21 vibrations taken to be active, provided that they assumed a collision efficiency of about 0.25 for the deenergization process. Lin and Laidler[4] repeated the calculations, using more reliable values[5] for the high-pressure Arrhenius parameters, and obtained agreement with experiment for both the geometrical and structural isomerization, without the use of low collision efficiencies. This agreement could be obtained with alternative models for the activated complex, and in general it is found that RRKM theory is not capable of distinguishing between mechanisms. Lin and Laidler also found that the

[1] E. W. Schlag and B. S. Rabinovitch, J. Amer. Chem. Soc., **82**:599 (1960).

[2] R. C. Golike and E. W. Schlag, J. Chem. Phys., **38**:1886 (1963).

[3] G. M. Wieder and R. A. Marcus, J. Chem. Phys., **37**:1835 (1962); G. M.Wieder, Ph.D. thesis, Polytechnic Institute of Brooklyn, 1961.

[4] M. C. Lin and K. J. Laidler, Trans. Faraday Soc., **64**:927, (1968).

[5] W. E. Falconer, T. F. Hunter, and A. F. Trotman-Dickenson, J. Chem. Soc., **1961**:609.

kinetic-isotope results of Rabinovitch et al.[1] were well interpreted by RRKM theory.

Decomposition of cyclobutane   The dissociation of cyclobutane into two ethylene molecules,

$$\begin{matrix} CH_2\!-\!CH_2 \\ |\qquad | \\ CH_2\!-\!CH_2 \end{matrix} \;\rightarrow\; \begin{matrix} H_2C\!=\!CH_2 \\ + \\ H_2C\!=\!CH_2 \end{matrix}$$

has been studied experimentally by Walters and coworkers,[2] who found the high-pressure rate constant to be given by

$$k^\infty = 10^{15.6} e^{-62,500/RT} \qquad sec^{-1}$$

The rate coefficient was found to fall off at low pressures.

Two distinctly different types of activated complex can be envisioned for this reaction.   The first involves the simultaneous lengthening of two of the carbon-carbon bonds and the contraction of the other two.

$$\begin{matrix} CH_2\!-\!CH_2 \\ |\qquad | \\ CH_2\!-\!CH_2 \end{matrix} \;\rightarrow\; \begin{matrix} H_2C\cdots CH_2 \\ \vdots\qquad\vdots \\ H_2C\cdots CH_2 \end{matrix} \;\rightarrow\; 2C_2H_4$$

In view of the very high frequency factor and entropy of activation for this reaction ($\Delta S^\ddagger \approx 9$ cal deg^{-1} mole^{-1}) it is necessary to assume that there is virtually free rotation of the ethylene molecules in the complex. The second mechanism involves the initial rupture of one carbon-carbon bond to give the tetramethylene biradical, the subsequent activated complex also having biradical character:

$$\begin{matrix} CH_2\!-\!CH_2 \\ |\qquad | \\ CH_2\!-\!CH_2 \end{matrix} \;\rightleftharpoons\; \begin{matrix} \qquad CH_2\!-\!CH_2 \\ \diagup\qquad\qquad\diagdown \\ H_2\!\cdot\!C\qquad\qquad\cdot CH_2 \end{matrix} \;\rightarrow$$

$$\begin{matrix} CH_2\cdots\cdots CH_2{}^\ddagger \\ \diagup\qquad\qquad\diagdown \\ H_2C\qquad\qquad CH_2 \end{matrix} \;\rightarrow\; 2C_2H_4$$

Wieder and Marcus[3] carried out RRKM calculations with the latter model, taking all vibrations as active and obtaining the correct high-

[1] B. S. Rabinovitch, D. W. Setser, and F. W. Schneider, *Can. J. Chem.*, **39**:2609 (1961).

[2] C. T. Genaux, F. Kern, and W. D. Walters, *J. Amer. Chem. Soc.*, **75**:6196 (1953); R. W. Carr and W. D. Walters, *J. Phys. Chem.*, **67**:1370 (1963); cf. H. O. Pritchard, R. G. Sowden, and A. F. Trotman-Dickenson, *Proc. Roy. Soc. (London)*, **A218**:416 (1953).

[3] G. M. Wieder and R. A. Marcus, *J. Chem. Phys.*, **37**:1835 (1962); G. M. Wieder, Ph.D. thesis, Polytechnic Institute of Brooklyn, 1961.

pressure frequency factor by loosening some of the vibrations. They were able to obtain good agreement with the experimental fall-off curves if they assumed a collision efficiency of 0.25. However, such a low value is unlikely; all the evidence indicates that collisions are very effective in transferring vibrational energy.[1] Lin and Laidler[2] pointed out that isomerizations have a special feature in that the newly formed product molecules may contain large amounts of vibrational energy, in contrast to decompositions in which much of the energy in the products may be in the form of translational energy. Before being deactivated an energy-rich molecule produced in an isomerization process may therefore easily change back into a reactant molecule. Lin and Laidler extended the RRKM theory to take account of this possibility and found that they could then obtain good agreement with the cyclobutane results with a collision efficiency of unity.[3] The treatment was also found to give a good interpretation of the kinetic isotope effects.

Calculations based on Slater's theory have also been made for this reaction,[4] but the energization rates are again calculated to be much smaller than the experimental values.

Isomerization of cyclobutene   The thermal isomerization of cyclobutene to butadiene,

$$\begin{array}{c} HC{=}CH \\ |\quad| \\ H_2C{-}CH_2 \end{array} \rightarrow H_2C{=}CH{-}CH{=}CH_2$$

has been studied by Walters and coworkers[5] and found to be a typical unimolecular reaction with a "normal" ($\sim 10^{13.4}$ sec^{-1}) frequency factor. Elliott and Frey[6] applied the RRKM theory to the results and found that to obtain good agreement, they had to postulate a collision efficiency of 0.3. Lin and Laidler[7] applied their extended RRKM theory, which allows for the back reaction, and were then able to obtain good agreement with a collision efficiency of unity.

[1] See H. S. Johnston, "Gas Phase Reaction Rate Theory," The Ronald Press Company, New York, 1966; G. H. Kohlmaier and B. S. Rabinovitch, *J. Chem. Phys.*, **38**:1692, 1709 (1963).

[2] M. C. Lin and K. J. Laidler, *Trans. Faraday Soc.*, **64**:94 (1968).

[3] M. C. Lin and K. J. Laidler, *Trans. Faraday Soc.*, **64**:927 (1968).

[4] D. Retzloff and J. Coull, *J. Chem. Phys.*, **47**:3927 (1967).

[5] W. Cooper and W. D. Walters, *J. Amer. Chem. Soc.*, **80**:4220 (1958); W. P. Hauser and W. D. Walters, *J. Phys. Chem.*, **67**:1328 (1963).

[6] C. S. Elliott and H. M. Frey, *Trans. Faraday Soc.*, **62**:895 (1966).

[7] M. C. Lin and K. J. Laidler, *Trans. Faraday Soc.*, **64**:94 (1968).

Isomerization of *cis*-butene-2   A kinetic study of the conversion of *cis*-butene-2 into the *trans* form,

$$
\begin{array}{ccc}
\ce{CH_3}\quad\ce{CH_3} & & \ce{CH_3}\quad\ce{H} \\
\diagdown\quad\diagup & & \diagdown\quad\diagup \\
\ce{C}=\ce{C} & \rightarrow & \ce{C}=\ce{C} \\
\diagup\quad\diagdown & & \diagup\quad\diagdown \\
\ce{H}\qquad\ce{H} & & \ce{H}\qquad\ce{CH_3}
\end{array}
$$

was made by Rabinovitch and Michel,[1] who observed a falling-off in the rate coefficient k^1 at pressures below 2 mm.   Wieder and Marcus[2] applied the RRKM theory to the results, but the agreement was not very satisfactory; an unreasonably low value, 0.02, had to be taken for the collision efficiency in order for the fall-off curves to be fitted.   Much better agreement was obtained by Lin and Laidler[3] with their modified RRKM theory, in which allowance was made for the back reaction.

Dissociation of ethane   The dissociation of ethane into two methyl radicals,

$$\ce{C2H6 -> 2CH3}$$

has been investigated by several workers,[4] and there is not complete agreement about the Arrhenius parameters; the equation

$$k_\infty = 3.2 \times 10^{16} e^{-88,000/RT} \qquad \sec^{-1}$$

is in reasonable agreement with all of the results.   The first-order rate coefficients fall off at pressures of a few millimeters.

The rather high frequency factor for the reaction, corresponding to an entropy of activation of about 13 cal mole^{-1} deg^{-1}, implies a rather loose structure for the activated complex.   Lin and Laidler[5] considered various possible models for the activated complex and made calculations based on RRKM theory:

1. Model 1 corresponds to "softened" vibrations; the frequencies of torsional and four bending motions were suitably reduced.

[1] B. S. Rabinovitch and K. W. Michel, *J. Amer. Chem. Soc.*, **81**:5065 (1959).

[2] M. C. Lin and K. J. Laidler, *Trans. Faraday Soc.*, **64**:94 (1968).

[3] *Ibid.*

[4] C. P. Quinn, *Proc. Roy. Soc. (London)*, **A275**:190 (1963); A. S. Gordon, *Symposium on Kinetics of Pyrolytic Reactions*, University of Ottawa, 1964; M. C. Lin and M. H. Back, *Can. J. Chem.*, **44**:505, 2357 (1966); A. B. Trenwith, *Trans. Faraday Soc.*, **62**:1538 (1966).

[5] M. C. Lin and K. J. Laidler, *Trans. Faraday Soc.*, **64**:79 (1968).

2. Model 2 is one in which the torsional motion has become a completely free rotation in the activated state.

3. Model 3 is the same as 2 except that the overall rotation of the molecule was taken to be *active* in the activated complex; this may be the case because of coupling of this rotation with the free internal rotation. The overall rotation in the *energized* molecule was assumed to be inactive.

Satisfactory agreement with experiment was obtained from all these models, especially 1 and 3, as is shown in Fig. 48. It is important to note in this connection that RRKM theory often does not permit discrimination between alternative models.

Decomposition of the ethyl radical The results[1] for the decomposition of the ethyl radical

$$C_2H_5 \rightarrow C_2H_4 + H$$

are consistent with the equation

$$k^\infty = 2.7 \times 10^{14}e^{-40,900/RT} \qquad sec^{-1}$$

[1] M. C. Lin and M. H. Back, *Can. J. Chem.*, **44**:2357 (1966); L. F. Loucks and K. J. Laidler, *Can. J. Chem.*, **45**:2795 (1967).

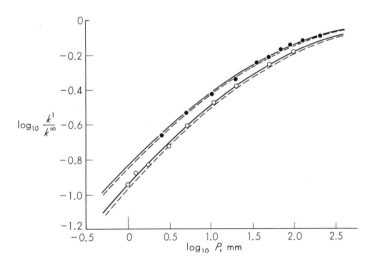

Fig. 48 Comparison of RRKM theory with experimental results for the ethane dissociation. The filled circles are experimental results at 913°K, the open circles at 978°K. The dashed lines show the results of calculations based on model 1, the solid lines based on model 3.

There is a substantial fall-off in the rate coefficient at a few millimeters pressure. The frequency factor is much closer to normal than for the ethane dissociation. Whereas in the latter reaction there is the breaking of a single bond, in the ethyl-radical decomposition there is in addition the formation of a double bond, and this no doubt produces a tighter structure in the activated state.

Lin and Laidler[1] considered two models for this reaction:

1. Model 1 involved no active rotation for either the energized radical or the activated complex.
2. Model 2 was one in which the torsional vibration was reduced to a free internal rotation in the energized species. Because of coupling of the overall rotation with this torsional rotation, both rotations were taken as active in the energized species. In the activated complex, on the other hand, the internal rotation is considered to have become a torsional vibration and only the overall rotation remained active.

Both models gave reasonable agreement with experiment.

It is to be seen from the few examples given that RRKM theory does provide a very satisfactory interpretation of the experimental results.

[1] M. C. Lin and K. J. Laidler, *Trans. Faraday Soc.*, **64**:79 (1968).

7
Molecular Dynamics

The preceding five chapters have been concerned with the theoretical treatments of rates on the basis of the equilibrium hypothesis. The present and the following chapter deal with theories in which this assumption is not made. It has been seen in Fig. 1 that theories may conveniently be classified according to whether or not they are explicitly related to potential-energy surfaces. The present chapter is concerned with theories of the movement of systems over potential-energy surfaces. This subject is conveniently referred to as *molecular dynamics*.[1] Theories which are not related to potential-energy surfaces are considered in Chap. 9 .

Some of the reasons why it is important for such nonequilibrium theories to be developed have been referred to in Chap. 1. We have seen that the equilibrium hypothesis is valid for reactions having substantial activation energies and which are therefore slow; it does, however, break down for very fast reactions since there will then be a depletion of the

---

[1] The term *kinematics* is also sometimes used, but this is to be avoided; kinematics refers only to properties of the motion that are independent of the nature of the forces.

more energized reactant species, with the result that the activated complexes cannot be at equilibrium with the reactants. Only by nonequilibrium theories can the equilibrium hypothesis be properly tested and the limits of its applicability established.

A second reason why nonequilibrium theories are being developed is that during recent years there has been increasing interest in what might be called the "fine structure" of chemical reactions. In particular, studies of infrared chemiluminescence and of reactions in crossed molecular beams have provided valuable information about the precise way in which reactive collisions occur and about the way in which the energy released in a reaction is distributed among the products. Many of the theoretical dynamical treatments are, in fact, being made with particular reference to the results of such experimental studies. An important objective of dynamical theories is the elucidation of the connection between the form of the potential-energy surface and the way in which the energy released passes into the products of reaction. An experimental study of energy distributions combined with a dynamical treatment for a variety of hypothetical potential-energy surfaces can lead to valuable conclusions about the actual shapes of potential-energy surfaces. Some examples of this procedure are considered later.

CLASSICAL DYNAMICS

Up to the present time most dynamical calculations have been made largely on the basis of classical mechanics. Usually quantum conditions have been imposed as far as the initial states are concerned, but the actual motion over the potential-energy barrier has been treated purely classically in the majority of studies. It is considerably easier to do this than to carry out a purely quantum-mechanical treatment, and it is important to consider what error may arise from making a purely classical calculation.

There are two main sources of error. The first relates to the fact that a purely classical treatment neglects the quantization, and therefore the existence of a zero-point level, at the activated state and at the other points along the reaction path. A classical calculation will, for example, permit a system to cross the activation barrier at an energy level which is lower than the zero-point level. As a result of this, for a given potential-energy surface a classical calculation will lead to a different activation energy and a different rate[1] than given by a quantum calculation or by using the activated-complex theory, which does take into account the

---

[1] A completely classical calculation would usually lead to a higher activation energy than the quantum calculation, because the zero-point energy is usually smaller for the activated complex than for the reactants.

quantization in the activated state. This matter is later considered in more detail for the H + H$_2$ reaction (pages 164 and 165).

The second source of error is that there may be a significant amount of quantum-mechanical tunneling through the barrier, an effect that is neglected in the classical calculations. This tunneling, as has been seen in Chap. 4, is unimportant if the potential-energy barrier is not too thin and if the energies of approach to it are not too small. If the barrier is thin or the energies of approach to the barrier are low, quantum-mechanical tunneling will make a significant contribution to reaction. It is difficult to give these statements a quantitative significance, since only a very limited amount of relevant computational work has been done and most of this has been related to rather simple and artificial potential-energy surfaces. Thus Hirschfelder and Wigner[1] treated the passage of a hydrogen atom over potential barriers of various kinds—abrupt, steep, and gradual. They found that the probability of crossing the barrier was a rapidly fluctuating function of the energy of the hydrogen atom. However, their calculations indicated that, under normal kinetic conditions, the distribution of energies would be sufficiently great to cover several periods of this fluctuation, with the result that the average rate of crossing the barrier corresponds closely to the classical value. Johnston[2] has also made calculations for hydrogen atoms crossing barriers of the London-Eyring-Polanyi-Sato (LEPS) type (see page 30), and concludes that there should be a large amount of quantum-mechanical tunneling at room temperature. As discussed earlier, however (see page 31), there is some reason to believe that the LEPS surfaces are unrealistically thin. More recently Karplus and coworkers[3] have made both classical and quantum-mechanical calculations for the same potential-energy surface. Their results, as will be discussed later (page 209), indicate a small but significant tunneling effect for the H + H$_2$ system. On the whole it appears that quantum-mechanical effects are not of great importance for hydrogen atoms, except perhaps at low temperatures; for reactions involving the transfer of heavier atoms it seems likely that classical treatments will be completely adequate as far as this aspect is concerned.

The first classical calculations of the motion over potential-energy barriers were made for the H + H$_2$ system by Eyring and Polanyi,[4] and further work was done by Hirschfelder, Eyring, and Topley.[5] The results

[1] J. O. Hirschfelder and E. Wigner, *J. Chem. Phys.*, **7**:616 (1939).

[2] H. S. Johnston, *Advan. Chem. Phys.*, **3**:131 (1960).

[3] M. Karplus, R. N. Porter, and R. D. Sharma, *J. Chem. Phys.*, **43**:3259 (1965); M. Karplus and K. T. Tang, *Discussions Faraday Soc.*, **44**:56 (1967).

[4] H. Eyring and M. Polanyi, *Z. Physik. Chem. (Leipzig)*, **B12**:279 (1931).

[5] J. O. Hirschfelder, H. Eyring, and B. Topley, *J. Chem. Phys.*, **4**:170 (1936).

have been reviewed[1] and will not be considered here since these pioneering calculations, important when they were done, were necessarily not very reliable. Computers had not then been developed, and point-by-point computations had to be made of the successive coordinates of the moving system. Because of the great amount of work involved the calculations could only be made for linear approaches of the H atom to the H_2 molecule, and the number of trajectories calculated was too small to make the results statistically meaningful.

During recent years, since the development of high-speed digital computers, several much more detailed calculations have been made. This work has already been of great value in leading to a number of important conclusions about the mechanisms of simple chemical reactions, even though in all cases there is doubt about the validity of the potential-energy surfaces used. It will be seen that one of the useful results of these calculations is that they open up the possibility of arriving at conclusions about the form of potential-energy surfaces by a comparison of dynamical calculations with experimental results.

The first computer calculations of reaction dynamics were performed by Wall, Hiller, and Mazur,[2] for the $H + H_2$ system. These calculations related to the LEP potential-energy surface, which has a basin, and integrations were carried out for the motion over the surface. The earlier calculations related to the linear $H\cdots H\cdots H$ complex, and the initial conditions were varied systematically. In the later work nonlinear complexes were considered also, but the motion was restricted to that in the plane of the H_3 complex; calculations of this type are referred to as 2D calculations. The starting conditions (rotational, vibrational, and translational energies, etc.) were chosen by a weighted random method. Unfortunately, in the second series of calculations very few collisions (6 out of 700) led to reaction, so that the results for the reactive collisions were not statistically significant. Wall and Porter[3] later carried out calculations using an empirical surface which had no basin,[4] but they also only considered linear complexes. Similar calculations were made by Blais and Bunker[5] and by Polanyi and Rosner.[6] They used a basinless LEPS surface, without the linearity restriction, but confined the motion

[1] S. Glasstone, K. J. Laidler, and H. Eyring, "The Theory of Rate Processes," McGraw-Hill Book Company, New York, 1941.

[2] F. T. Wall, L. A. Hiller, and J. Mazur, *J. Chem. Phys.*, **29**:255 (1958); **35**:1284 (1961).

[3] F. T. Wall and R. N. Porter, *J. Chem. Phys.*, **39**:3112 (1963).

[4] F. T. Wall and R. N. Porter, *J. Chem. Phys.*, **36**:3256 (1962).

[5] H. C. Blais and D. L. Bunker, *J. Chem. Phys.*, **37**:2713 (1962).

[6] J. C. Polanyi and S. D. Rosner, *J. Chem. Phys.*, **38**:1028 (1963).

of the atoms to a single plane (2D). These calculations were the first to provide an indication that when the collisions satisfy certain requirements as to energy and angles of approach, the system passes directly through the activated state; the activated complex, in other words, has such a short lifetime that it does not perform a complete vibration or rotation. A potential-energy surface with a basin tends to lead to a complex of longer life; since molecular-beam experiments usually indicate a complex of short life, it seems likely that basins are unusual. This matter is referred to again later in much greater detail.

Much more extensive classical dynamical calculations have more recently been made by several groups of workers, particularly Bunker, Karplus, and J. C. Polanyi and their coworkers. These calculations have considerably more statistical significance than the earlier ones and will be considered in some detail in the present chapter. The general methods employed will first be described.

The dynamical equations The first step in a dynamical study is to calculate the potential-energy surface. The methods for doing this have been outlined in Chap. 2. For the calculations to be as realistic as possible most of the work has been done with surfaces that have been arrived at with the use of a good deal of empiricism.

The next step is to write down the classical hamiltonian for the system. For the reaction

$$A + BC \rightarrow AB + C$$

the classical hamiltonian has the form

$$H = (2\mu_{B,C})^{-1} \sum_{j=1}^{3} P_j^2 + (2\mu_{A,BC})^{-1} \sum_{j=4}^{6} P_j^2 + V(Q_1, Q_2, \ldots, Q_6)$$

$$(1)$$

Here Q_j ($j = 1, 2, 3$) represents the cartesian coordinates of particle B with respect to C, and Q_j ($j = 4, 5, 6$) represents the cartesian coordinates of particle A with respect to the center of mass of BC. The P_j ($j = 1, 2, \ldots, 6$) are the momenta conjugate to the Q_j, and $V(Q_1, Q_2, \ldots, Q_6)$ is the potential-energy function. The μ's are the reduced masses defined by

$$\mu_{B,C} = \frac{m_B m_C}{m_C + m_B} \tag{2}$$

$$\mu_{A,BC} = \frac{m_A(m_B + m_C)}{m_A + m_B + m_C} \tag{3}$$

Hamilton's equations for this system are

$$\dot{Q}_j = \frac{\partial H}{\partial P_j} \qquad j = 1, 2, \ldots, 6 \tag{4}$$

$$\dot{P}_j = -\frac{\partial H}{\partial Q_j} = -\frac{\partial V}{\partial Q_j} \tag{5}$$

Equations (4) and (5) comprise a set of 12 simultaneous first-order differential equations; integration of them, subject to a complete set of initial conditions, leads to the collision trajectories, i.e., to a description of the motion of the system over the surface.

Monte Carlo averaging procedure   For the results of the calculations to have statistical significance it is necessary to choose a representative group of initial conditions by the use of an averaging procedure.   The most commonly used is the *Monte Carlo* procedure, which is a method of making a weighted random selection of the initial parameters.

The initial parameters that have to be considered are:

1. The velocities of the reactant molecules relative to the center of mass of the system.   Thus for a reaction A + BC → AB + C an initial parameter is the relative velocity of A with respect to the center of mass of BC.
2. The vibrational energy of the reactant molecules.
3. The vibrational phase of the reactant molecules relative to their approach.
4. The rotational energy of the reactants.
5. The rotational phase of the reactants.
6. The impact parameter b.   This, as illustrated in Fig. 49, is the closest distance of approach of A to the center of mass of BC if the two molecules continued with their initial velocities without interacting with each other.

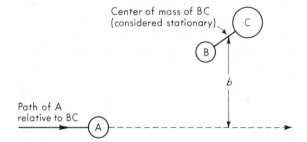

Fig. 49   The impact parameter b.

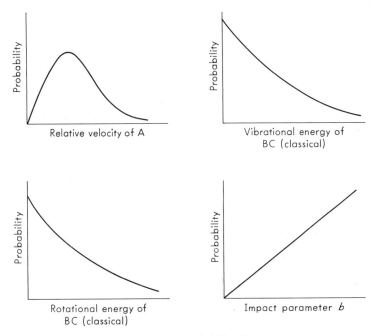

Fig. 50   The form of the various probability functions.

The weighting of these initial parameters is necessary because associated with each variable there is a probability distribution function. Some examples of such functions are shown schematically in Fig. 50. Details of how the Monte Carlo weighting procedure is applied are given by Blais and Bunker[1].

After a complete set of initial conditions has been arrived at, Eqs. (4) and (5) can be integrated numerically to yield the collision trajectory, i.e., the path over the potential-energy surface.   This procedure must be repeated a large number of times, and statistical averaging performed. To obtain statistically significant results, a thousand or more trajectories, corresponding to reactive collisions, should be calculated.   The trajectories calculated for a large number of initial conditions then provide the required information about the detailed nature of the reactions.   Rate constants can be calculated, and it has also been found convenient to express results in terms of *cross sections*, which are equivalent to the cross sections based on the hard-sphere theory of reactions.   In addition to cross sections, trajectory calculations yield important information about the most probable paths of the reaction products; this, as will be seen,

[1] Blais and Bunker, *loc. cit.*

is relevant to the results of infrared chemiluminescence and molecular-beam experiments.

THE REACTION $H + H_2 \rightarrow H_2 + H$

The dynamical calculations of Karplus, Porter, and Sharma[1] on the $H + H_2$ reaction represent a very important contribution to our understanding of the details of a chemical reaction. The main results of this work are outlined in the present section.

The potential-energy surface used was the semiempirical one developed by Porter and Karplus[2] and shown in Fig. 14; it has no basin. The dynamical calculations were semiclassical; the initial states of the H_2 molecule were weighted according to quantum-mechanical partition functions, but the rest of the problem was treated classically. That is, the zero-point level and the quantized vibrational and rotational levels of the H_2 molecule were explicitly considered, and the statistical distribution of the molecules over these levels was taken into account. However, no zero-point level or other quantized levels were assigned to the activated state. The consequences of this are discussed later.

Method of calculation   The general computational procedure was that described earlier (page 157); the hamiltonian was set up with respect to the potential-energy surface and the equations of motion solved. The following initial conditions were explicitly considered:

1. The rotational quantum number J of the H_2 molecule.
2. The vibrational quantum number v of the H_2 molecule.
3. The initial relative velocity V_R of the atom H with respect to the molecule H_2.
4. The impact parameter b (see Fig. 49).
5. The initial distance ρ between H and the center of mass of H_2. This distance is chosen quite arbitrarily; it is made fairly small to save computing time, but must be sufficiently large for there to be essentially no interaction between H and H_2.
6. The initial orientation of the molecule H_2 to H. This is specified by spherical polar coordinates R, θ, and ϕ.
7. The internal momentum of the molecule H_2. This is specified by the angle η of the momentum vector relative to an arbitrarily chosen vector that is perpendicular to the molecular axis.

[1] M. Karplus, R. N. Porter, and R. D. Sharma, *J. Chem. Phys.*, **34**:3259 (1965).

[2] R. N. Porter and M. Karplus, *J. Chem. Phys.*, **40**:1105 (1964).

The initial state of a trajectory is determined completely by these nine variables J, v, V_R, b, ρ, R, θ, ϕ, and η, and various procedures can be used for assigning values to them. In most of the calculations the variables R, θ, ϕ, and η are selected at random from appropriate distribution functions, and enough trajectories are computed to permit Monte Carlo averaging over their values. The quantity ρ, as previously indicated, is chosen quite arbitrarily. In the calculations of the reaction probability P_r the initial variables J, v, V_r, and b are assigned a number of values, an important part of the investigation being to see how P_r depends on one of these variables, with the other three held constant. The reaction cross section S_r is obtained by integrating over the impact parameter b [see Eq. (8) below], and the calculations therefore reveal how this quantity depends on the initial parameters J, v, and V_r. The rate constant k is obtained by considering the Boltzmann distribution of J, v, and V_r at the appropriate temperature, and it is derived by a summation procedure.

The reaction probability   If the effects of the initial parameters R, θ, ϕ, and η are averaged out by a Monte Carlo procedure, it is possible to fix V_R, J, v, and b and determine the reaction probability P_r. This is done by first determining N_r, the number of reactions that result from a set of N trajectories with the initial values for R, θ, ϕ, and η chosen at random from a suitable set of distribution functions. The reaction probability is then given by

$$P_r(V_R,J,v,b) = \lim_{N \to \infty} \frac{N_r(V_R,J,v,b)}{N(V_R,J,v,b)} \tag{6}$$

A study was made of the way P_r varies with the impact parameter b, with V_R, J, and v given fixed values. Figure 51 shows the results

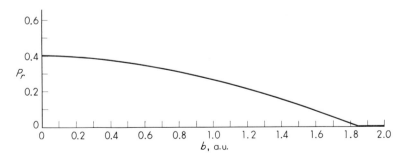

Fig. 51   The variation of the reaction probability P_r with the impact parameter, for $V_R = 1.17 \times 10^6$ cm/sec, $J = 0$, and $v = 0$.

with $V_R = 1.17 \times 10^6$ cm/sec, $J = 0$, and $v = 0$. It was found that the dependence of P_r on b is represented satisfactorily by the following function

$$P_r = a \cos \frac{\pi b}{2b_m} \qquad b \leq b_m$$

$$P_r = 0 \qquad\qquad b > b_m$$

For $J = v = 0$ and $V_R = 1.17 \times 10^6$ cm/sec the values of a and b_m are seen from the graph to be given by

$$a \approx 0.39 \qquad b_m = 1.85 \text{ a.u.}$$

The reaction probability shows the expected type of variation with b, reaction becoming highly unlikely at impact parameters that are too large. It is to be noted that the reaction probability does not approach unity for an impact parameter of zero. This is explained by a *steric* requirement for reaction. A simple way of looking at the situation is to say that for $b = 0$ there is unit probability of reaction if the angle between the coordinate of the atom H relative to the center of mass of H_2 is less than a certain angle β_0, and that the probability is zero if the angle is greater than β_0. It must be emphasized that this concept is oversimplified, since the forces on the molecule in fact tend to line it up with the incoming atom.

The reaction cross section   If, instead of varying in the manner shown in Fig. 51, the reaction probability were unity up to b_{max} and zero at larger values of b, the reaction cross section would be πb_{max}^2. At all b values, however, P_r is less than unity, and the cross section (see Chap. 1) is instead given by

$$S_r = \int_0^{b_{max}} P_r \, d(\pi b^2) \tag{7}$$

the areas πb^2 having been weighted by the probability factor P_r. This equation can be written as

$$S_r(V_R, J, v) = 2\pi \int_0^{b_{max}} P_r(V_R, J, v, b) b \, db \tag{8}$$

On the basis of calculations for a large number of reaction trajectories Karplus and coworkers were able to study the dependence of the reaction cross section S_r on the relative velocity V_R and the quantum numbers J and v. Figure 52 shows the variation of S_r with V_r for $v = 0$ and $J = 0$; similar figures were constructed for other v and J combinations. Figure 53 shows a similar plot, also for $v = 0$ and $J = 0$, of S_r against the relative energy E_R, which is the kinetic energy corresponding to the relative velocity V_R. The calculations show that the energy

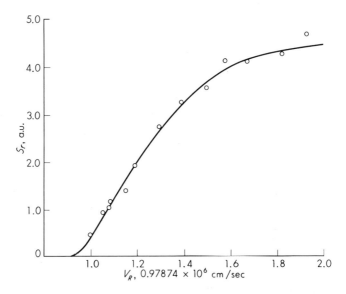

Fig. 52 Reaction cross section $S_r(V_R,J,v)$ as a function of V_R for H + H₂; $v = 0$ and $J = 0$.

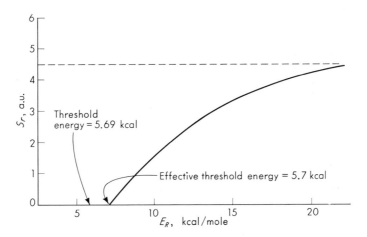

Fig. 53 Variation of the reaction cross section S_r with the relative energy of approach E_R, for collisions between H atoms and H₂ molecules in their lowest vibrational and rotational states ($v = 0$ and $J = 0$).

threshold $E_R{}^0$, below which S_r is zero, is 5.69 kcal/mole. However, it is to be seen from Fig. 53 that S_r remains very close to zero up to a value of about 7 kcal and then rises steadily to an asymptotic value.

Comparison of the S_r versus V_R curves for different J values shows that the dependence on J is not very great; that is, curves of the type in Fig. 52 for different J values are very similar to one another. The threshold values tend to become *larger* as J is increased. The variations with the vibrational quantum number v have not been treated in as much detail, since in the usual temperature range most of the H_2 molecules have v equal to zero. The calculations do indicate, as expected, that the threshold energy is *decreased* markedly as the v value increases.

Threshold energy values   The barrier height of the potential-energy surface is 9.13 kcal with respect to the classical ground state of $H + H_2$, and the zero-point vibrational energy of the initial state ($v = 0$, $J = 0$) is 6.20 kcal. The minimum relative translational energy required to cross the barrier from the zero-point level is therefore 2.93 kcal. These and other energy relationships are shown in Fig. 54. The computed threshold energy (see Fig. 53) is, however, much greater than this, namely, 5.69 kcal. It follows that not all the zero-point energy and relative translational energy is available for crossing the barrier; some 2.76 kcal of addi-

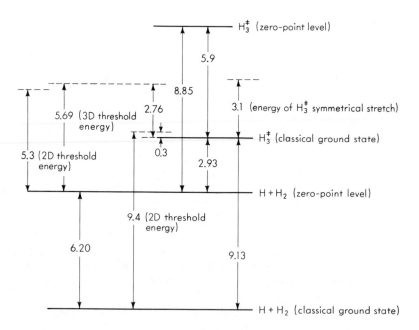

Fig. 54   Energy diagram for the $H + H_2$ system.

tional energy is needed, and this will be present as vibrational energy in the activated state. This energy is not far from the 3.1 kcal of energy involved in the symmetrical vibration of the linear $H_3^\ddagger$ complex. It is important to realize that the difference between the 2.93 kcal required to reach the top of the barrier and the 5.69 kcal that the system $H + H_2$ must have in relative translational energy is *not* due to the zero-point energy of the activated state; since the trajectories are calculated classically this zero-point energy is not taken into account.

Some light on the reason for the difference is shed by calculations in which the H_2 molecule is initially in its classical ground state, so that it requires 9.13 kcal to reach the top of the barrier. The calculations now show that for a linear activated complex a relative translational energy of at least 9.4 kcal is required for the system to cross the barrier; this corresponds to an energy of only about 0.3 kcal in excess of the barrier height. This small difference is due to the curvature of the reaction path. If the same calculation for a linear activated complex is done for H_2 molecules with zero-point energy, the threshold translational energy is found to be 5.3 kcal; that is, of the additional 6.20 kcal of vibrational energy in the system, about 4 kcal can contribute to reaction. This means that there is some degree of *adiabaticity* in the vibrational degrees of freedom; not all of the energy of vibration in the H_2 molecule is available for crossing the barrier, but some is forced into the symmetric stretching vibration of the activated complex.

Calculations of this kind were also made for the three-dimensional (3D) case. It might be expected that the threshold values would be the same as for the linear case, since the linear activated complex is the one of lowest energy. However, because of the low probability of linear collisions the *effective* threshold energy is 5.7 kcal, rather than the 5.3 kcal found in the linear case. The situation in the 3D case is much more complicated than in the linear case, since energy can now pass into the bending motions of the complex. One consequence of this is that there is a relatively slow increase of cross section with E_R in the neighborhood of the threshold energy; much of the energy can be "wasted" in symmetrical and bending vibrations of the complex, so that only highly restricted initial conditions permit reaction near the threshold.

That the threshold values tend to become larger as J increases (see page 164) is related to this effect. A high degree of rotation in the H_2 molecule leads to a decrease in the orienting effect of the interaction potential between H and H_2, an effect that favors the linear activated complex, which is the one of lowest energy.

Rate constants and activation energy From a knowledge of P_r and S_r, the reaction rate can be calculated by summation or integration over the distribution functions for V_R, J, and v that are characteristic of the system under consideration.   For the temperature range of experimental interest (300 to 1000°K) only the $v = 0$, 1, 2, 3, 4, and 5 states of the hydrogen molecule make a significant contribution to the equilibrium population.

Table 6 gives the calculated rate constants from 300 to 1000°K, together with some experimental values about which there is some uncertainty.   The agreement is seen to be quite satisfactory, although this may be accidental since the potential-energy surface used is of unknown reliability.   Also, no tunneling corrections have been employed, and these might substantially raise the rate constants at the lower temperatures.

Figure 55 shows an Arrhenius plot of the calculated rate constants.   The Arrhenius law is satisfactorily obeyed, the kinetic parameters being

$$A = 4.33 \times 10^{13} \text{ cc mole}^{-1} \text{ sec}^{-1} \qquad E = 7.44 \text{ kcal/mole}$$

These values are in satisfactory agreement with what appear to be the best experimental estimates,[1] i.e.,

$$A = 5.4 \times 10^{13} \text{ cc mole}^{-1} \text{ sec}^{-1} \qquad E = 7.5 \pm 1 \text{ kcal/mole}$$

There is a slight deviation from strict Arrhenius behavior, and this can be traced to the influence of the rotational quantum number J on the energy threshold values.   An equation which gives a slightly better

[1] A. Farkas and L. Farkas, *Proc. Roy. Soc. (London)*, **A152**:124 (1935); G. Boato, G. Careri, A. Cimoni, E. Molinari, and G. G. Volpi, *J. Chem. Phys.*, **42**:783 (1956).

Table 6   Calculated and observed rate constants

Temperature, °K	Calculated $k \times 10^{-11}$, cc mole^{-1} sec^{-1}	Observed† $k \times 10^{-11}$, cc mole^{-1} sec^{-1}
300	0.00185	0.0014–0.0020
400	0.0356	0.017–0.061
500	0.22	
600	0.780	
700	1.97	2.49–4.99
800	4.05	4.66–9.32
900	7.20	7.6–15.2
1000	11.53	11.0–22.0

† References to observed values are given in the paper by Karplus, Porter, and Sharma, *J. Chem. Phys.*, **34**:3259 (1965).

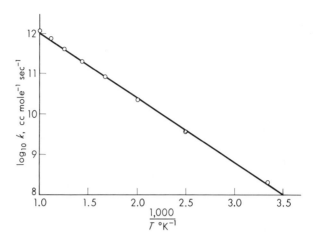

Fig. 55 Arrhenius plot for the calculated dynamical rate constants for the $H + H_2$ reaction.

fit than the simple Arrhenius equation is

$$k = 7.87 \times 10^9 T^{1.18} e^{-6,230/RT}$$

Comparison with activated-complex theory It is of great interest to compare the results of the dynamical calculations with those based on activated-complex theory for the same potential-energy surface. At the outset, it should be noted that there is an important difference between the two methods of approach, which makes it difficult to make a completely satisfactory comparison between the two procedures. In the dynamical calculations only the initial states are treated quantum mechanically; the trajectories allow for no quantization during the course of reaction. In the conventional activated-complex theory, on the other hand, both initial and activated states are treated quantum mechanically. Karplus et al. made the activated-complex calculations in the conventional way, with the activated complex treated quantum mechanically, even though this involves a different assumption from that used in the dynamical calculations. Their results are shown in Table 7. The activated-complex rates are seen to be lower than the dynamical ones by a factor of about 7 at $300°K$ and by a factor of about 1.25 at $1000°K$. The activated-complex theory leads to the rate equation

$$k = 7.4 \times 10^{13} e^{-8,800/RT} \qquad \text{cc mole}^{-1} \text{ sec}^{-1}$$

as compared with

$$k = 4.3 \times 10^{13} e^{-7,400/RT} \qquad \text{cc mole}^{-1} \text{ sec}^{-1}$$

**Table 7   Rate constants obtained from the dynamical
calculations and from activated-complex theory**

Temperature, °K	Rate constant $\times 10^{-11}$	
	Dynamical treatment	Activated-complex theory
300	0.00185	0.000301
400	0.0356	0.0111
500	0.221	0.098
600	0.780	0.430
700	1.97	1.26
800	4.05	2.88
900	7.20	5.58
1000	11.53	9.64

for the dynamical calculations.   The activated-complex activation energy
of 8.80 kcal/mole is very close to the difference between the zero-point
levels in the initial and activated states, which is 8.85 kcal/mole (see Fig.
54); the latter would be the hypothetical activation energy of the absolute
zero, and the effect of temperature is seen to be small.

It is evident that the reason why the dynamical activation energy is
lower is that quantization in the activated state is not taken into account.
On the dynamical model the crossing of the barrier is limited only by classical adiabatic restraints, as discussed earlier; the system can cross the
barrier at an energy lower than the zero-point level.   This explains the
lower activation energies and higher rates given by classical dynamical
calculations.

Later in this chapter (pages 206 to 209) will be considered some
of the purely quantum-mechanical calculations.   These are much more
difficult to carry out than the classical ones and so far have not
revealed anything like as much detail as have the classical calculations.
It might be worth exploring classical calculations that in some way introduce quantum restrictions at the activated state.   For example, it would
be interesting to explore the classical motion over a potential-energy surface on which had been placed a "floor" corresponding to the zero-point
level at each point along the reaction path.

Collision details   Direct information about the $H + H_2$ collisions is provided by a study of trajectories.   Figure 56 shows some typical trajectories for nonreactive collisions, and Fig. 57, for reactive collisions.   It is
to be seen from Fig. 56 that there is relatively little energy exchange during the nonreactive collisions, the amplitude of vibration of the molecule

H^β—H^γ being little affected by the collision with H^α; the collisions are largely *elastic*. A little more energy exchange occurs as a rule for reactive collisions; in Figs. 57*a* and 57*b* the vibrational amplitude of the new molecule H^α—H^β is significantly different from that of the original molecule H^β—H^γ. The reactive collisions, however, are largely elastic, the energy exchange never being very considerable.

Two limiting models have been used for exchange reactions of this type. The first of these is the *long-lived collision-complex* or *compound-state* model, in which the colliding molecules form a complex whose lifetime is long compared with the periods of its rotation and vibration. In this situation the initial energy of relative motion has time to become distributed among the degrees of freedom of the complex, which subsequently decomposes in a manner that is independent of the way it was formed; in decomposing the complex has no "memory" of how it was formed. The alternative model is the *simple collision* or *direct-interaction*

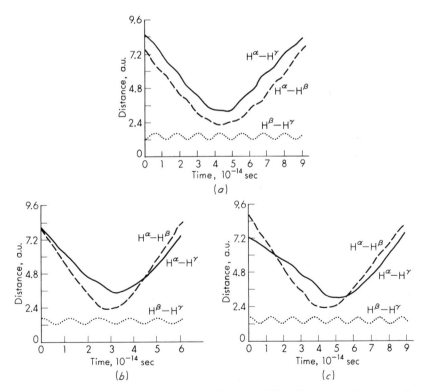

Fig. 56 Typical collision trajectories for the H^α + H^β − H^γ system, for unreactive collisions. The graphs show the variations with time of the H^α—H^β, H^β—H^γ, and H^α—H^γ distances.

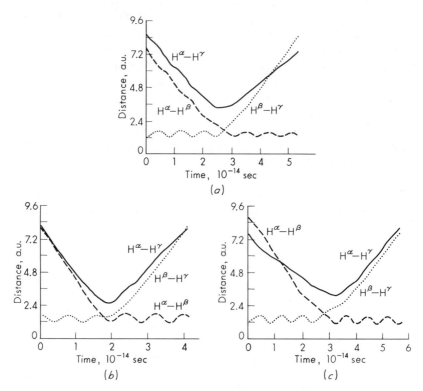

Fig. 57 Corresponding plots to those in Fig. 56 for typical reactive collisions.

model, in which the collision time is of the order of the time required for the reactant molecules to pass one another if there were no interaction. The lifetime is now shorter than the period of vibration or rotation of the complex, and the decomposition of the complex depends on the manner in which it was formed. As has been seen, this second model is the one assumed in activated-complex theory.

Figures 56 and 57 show clearly that the direct-interaction model is the one favored by the dynamical calculations. For both reactive and nonreactive collisions the complex has a lifetime of only about 10^{-14} sec. Since the vibrational period of H_2 is about 0.5×10^{-14} sec and the rotational period is larger ($\sim 2 \times 10^{-13}$ sec for $J = 1$) the collisional interaction time is much too short to permit equilibration of energy among the various degrees of freedom of the system. The calculations did reveal a very small proportion of collisions in which the complex underwent a complete vibration or two, but this is very atypical behavior.

It is important to note that the result that the reaction proceeds mainly by direct interaction is quite consistent with activated-complex theory, in spite of the fact that activated-complex theory treats the activated complexes as existing at equilibrium concentrations with the reactants. This situation is not to be regarded as arising because the complexes have time to come to equilibrium; on the contrary, according to the theory they are created in a state of equilibrium from the reactants.

CHEMILUMINESCENCE

In view of the difficulty that exists in arriving at reliable potential-energy surfaces, it is important to be able to draw conclusions about the general form of surfaces from experimental results. Overall kinetic studies do not contribute much information of this kind, and it is necessary to turn to rather special techniques which are more useful in providing information about the details of reaction processes. The most important of these are chemiluminescence studies and molecular-beam investigations.

Important results relating to the energetics of chemical reactions have been obtained for reactions between alkali metals and halogens. M. Polanyi[1] initiated work of this type by studying the emission of light when sodium reacts with halogens. His work allowed him to arrive at rate constants for reactions such as

$$Na + Cl_2 \rightarrow NaCl + Cl$$

and

$$Na_2 + Cl \rightarrow Na + NaCl$$

In the second reaction a substantial fraction of the energy released remained as vibrational energy in the NaCl molecule, which in a subsequent collision with a sodium atom gave rise to electronically excited Na, which emitted the sodium D line,

$$NaCl' + Na \rightarrow NaCl + Na^*$$

The amounts of energy released when Na_2 reacts with halogen atoms are as indicated below:

$$Na_2 + Cl \rightarrow NaCl + Na + 80\,kcal$$
$$Na_2 + Br \rightarrow NaBr + Na + 69\,kcal$$
$$Na_2 + I \rightarrow NaI + Na + 55\,kcal$$

In each case at least 48 kcal/mole must reside in the resulting sodium halide molecule, since this is the energy required to excite the Na atom in

[1] M. Polanyi, "Atomic Reactions," Williams and Norgate, London, 1932; confirmation that chemiluminescence arises from the sort of energy exchange postulated by Polanyi has been obtained in the molecular-beam experiments of M. C. Moulton and D. R. Herschbach, *J. Chem. Phys.*, **44**:3010 (1966).

the subsequent step.   Similar results were obtained[1] in reactions of potassium, where the corresponding energies are

$$K_2 + Cl \rightarrow KCl + K + 43 \text{ kcal}$$
$$K_2 + Br \rightarrow KBr + K + 45 \text{ kcal}$$
$$K_2 + I \rightarrow KI + K + 40 \text{ kcal}$$

The energy required to excite K electronically is 37 kcal/mole; intense chemiluminescence was observed in these systems, which indicated that at least this amount of energy was produced as vibrational energy in the potassium halide molecule.

Harpooning   An interesting feature of the reactions of the type $Na_2 + X$ and $K_2 + X$ is that the collision cross sections are rather large, about 50 to 100 Å^2, so that reaction can occur when the reactants are 2 to 3 Å apart.   The same is true for the reactions

$$Na + X_2 \rightarrow NaX + X$$

Polanyi explained these results in terms of an ionic intermediate $A^+\cdots B^-\cdots C$; thus

$$A + B—C \rightarrow A^+\cdots B^-\cdots C \rightarrow A—B + C$$

The atom A has been described as "tossing out its valence electron to hook atom B, which it hauls in with the coulomb force of attraction"; Polanyi referred to the mechanism as an *electron-jump* or *harpooning* mechanism.   Also, to explain why such reactions lead to vibrational excitation in the product molecule B—C, Evans and Polanyi[2] suggested that the potential-energy surface is of a particular kind later[3] described as an *attractive* one; the term *early downhill* has also been used to describe such surfaces.   A surface of an extremely attractive type is shown in Fig. 58. The essential feature of an attractive surface for a reaction A + BC is that the potential-energy decrease after the activated state has been attained should occur while the A$\cdots$B bond is changing its length from an extended one to the normal A—B distance; consequently the heat of reaction is released when the A—B bond is extended, so that a considerable amount of energy passes into vibrational energy of the A—B bond.

[1] M. Krocsak and G. Schay, *Z. Physik. Chem. (Leipzig)*, **B19**:344 (1932); E. Roth and G. Schay, *Z. Physik. Chem. (Leipzig)* **B28**:323 (1935).

[2] M. G. Evans and M. Polanyi, *Trans. Faraday Soc.*, **31**:875 (1935); **35**:178 (1939).

[3] J. C. Polanyi, in R. Stoops (ed.), "Transfert d'Energie dans les Gaz," pp. 177, 526, Interscience Publishers, Inc., New York 1962; *J. Quant. Spectr. Radiative Transfer* **3**:471 (1963).

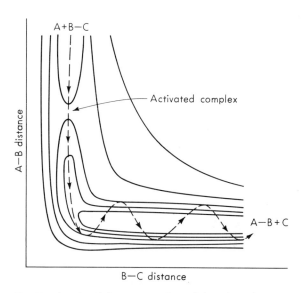

Fig. 58 A potential-energy surface of the attractive type, for the reaction $A + BC \rightarrow AB + C$. The dashed line shows a typical reaction path, in which much of the energy released passes into vibration of the AB molecule.

Potential-energy surfaces Magee[1] made a theoretical study of these alkali-metal–halogen reactions, in terms of potential-energy surfaces. By including surfaces for ionic states he was able to interpret the large cross sections, which arise from the crossing of ionic and covalent surfaces at high separations. The main points in Magee's treatment may be seen from Fig. 59, which shows the potential-energy curves for some of the lowest electronic states of an alkali halide molecule. In the ground state the molecule is essentially M^+X^- near its equilibrium bond distance, but its dissociation gives the atoms M and X and not the ions M^+ and X^-. The zeroth-order potential-energy curve for the ionic state must therefore cross that for the covalent state, as shown in the figure. These configurations interact to some extent so that the curves do not cross; if the configuration mixing is weak, the curves will approach very closely. For all the alkali halides the $M \cdots X$ separation r_c at the crossing point is large, the separation being ~ 15 Å for $K \cdots Br$. At this separation the coulombic attraction is dominant, and r_c may therefore be calculated from the energy

[1] J. L. Magee, *J. Chem. Phys.*, **8**:687 (1940); see also S. Glasstone, K. J. Laidler, and H. Eyring, "The Theory of Rate Processes," McGraw-Hill Book Company, New York, 1941.

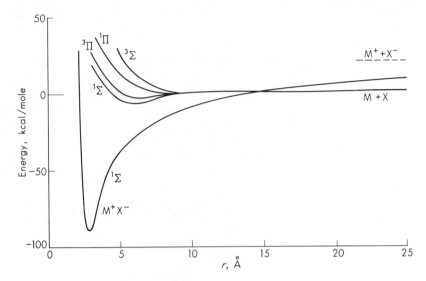

Fig. 59 Potential-energy curves for the KBr molecule, showing the crossing of ionic and covalent states.

required to form the ion pair by using the relationship

$$\frac{e^2}{r_c} \approx I(M) - E(X) \tag{9}$$

Here $I(M)$ is the ionization potential of the alkali atom M, and $E(X)$, the electron affinity of the halogen atom X.

For the lowest state of the M-X system the electron distribution thus undergoes an abrupt change in the vicinity of r_c; if M and X atoms are brought together, there may be, at the distance r_c, a jump of the electron from M to X. In fact, however, for most of the alkali halides, the ionic-covalent configuration mixing is very weak at the large r_c separations involved,[1] so that the electron has little chance of jumping as the region of the crossing point is traversed (this is shown by the optical spectra of the halides).

In a reaction of the type M + X—Y Magee's treatment shows that the situation is essentially the same. The potential-energy surfaces for the M····X····Y and M$^+$····X$^-$····Y configurations must cross, and the electron jump occurs at the crossing point. The situation is, however, somewhat different from the M····X case, since the electron switch now occurs with high probability. The value of r_c for the M-X-Y system is

[1] R. S. Berry, *J. Chem. Phys.*, **27**:1288 (1957).

given by

$$\frac{e^2}{r_c} \approx I(\mathrm{M}) - E^v(\mathrm{XY}) \tag{10}$$

where $E^v(\mathrm{XY})$ represents the vertical[1] electron affinity of the XY molecule. Usually $E^v(\mathrm{XY})$ will be smaller than the affinity of the X atom, so that the r_c values will be considerably smaller for the M-X-Y system than for the M-X system.

On the basis of these concepts Herschbach[2] has constructed the diagram shown in Fig. 60 for the K $+$ Br$_2$ system. For large separations the potential energy corresponds to attraction between K and Br$_2$ as a

[1] That is, the electron affinity corresponding to the formation of XY$^-$ having the same X—Y distance as in XY.

[2] D. R. Herschbach, in J. Ross (ed.), "Molecular Beams," p. 371, Interscience Publishers, Inc., New York, 1966.

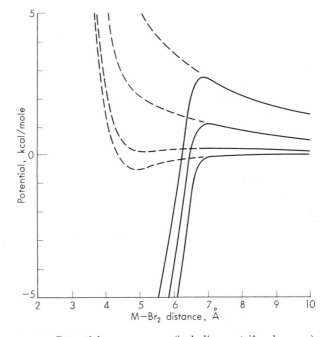

Fig. 60 Potential-energy curves (including centrifugal energy) for the collision of K and Br$_2$. The dashed curves include the dispersion and repulsion terms; the solid curves show the energies when the harpoon effect is included. Three sets of curves are shown, corresponding to different amounts of collisional angular momentum.

result of dispersion forces, and the dashed curves in the diagram correspond to these forces together with repulsive forces, which become important at small distances.   The solid lines show the modifications that arise
when account is taken of the harpooning effect of the electron switch.
As the K and Br_2 approach one another they interact to only a very small
extent, and this situation persists almost until the separation r_c is reached;
at this point the arrival of the harpooning electron produces a transition
between the potential-energy curves for the isolated Br_2 and Br_2^- molecules, as illustrated in Fig. 61.   The Br_2^- curve shown in this figure is
severely distorted by the strong electric field of the approaching K^+ ion,
and the complex dissociates into K^+Br^- and Br.   The Br_2^- ion is seen to
be forced into a highly excited vibrational state, barely below the dissociation limit, and dissociation will certainly occur as a result of the presence
of the K^+ ion.

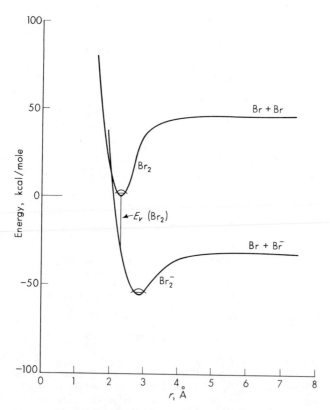

Fig. 61   Potential-energy curves for the ground electronic state
of Br_2 and its negative ion Br_2^-.

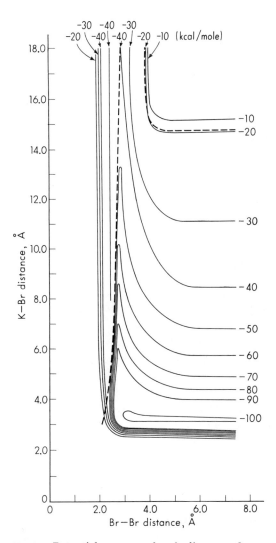

Fig. 62 Potential-energy surface for linear configurations of the K-Br-Br system. (*After Herschbach.*)

There can be an appreciable lag between the arrival of the harpooning electron and that of M^+. The time required for the breakup of Br_2^- is less than the vibrational period, or $<10^{-13}$ sec. The time taken for K^+ and Br_2^-, starting at rest at 7 Å apart, to reach the K^+Br^- equilibrium distance ($\sim$3 Å) is about 10^{-13} sec. Thus, when the K^+ ion arrives at the distance of 3 Å from the Br^- ion, the Br atom may have already departed.

Figure 62 shows the complete potential-energy surface for the

K + Br$_2$ reaction.[1]   It is clearly of the attractive type, as is shown by comparison with Fig. 59.   The way in which this type of potential-energy surface explains the molecular-beam results is shown later in this chapter.

Other reactions   A considerable number of other reactions have been studied spectroscopically, with the object of determining the extent to which the energy passes into vibration of the product molecules.   For the reaction

$$H + Cl_2 \rightarrow HCl + Cl + 45\ kcal$$

the evidence[2] indicates that about 55 percent of the energy released passes into translation and rotation of the products, with only a small amount passing into vibration.   This situation is therefore very different from that found in the alkali-metal reactions.

The explanation suggested for this contrasting behavior is that the hydrogen-atom reactions occur, not on an attractive surface of the type shown in Fig. 58, but on a *repulsive* or *late-downhill* surface.   In the case of a repulsive surface, an extreme example of which is shown schematically in Fig. 63, the heat of reaction is liberated along the coordinate which corresponds to increasing separation of the products AB + C.   A typical reaction path is shown, and there is seen to be little vibrational energy in the product AB, most of the heat of reaction going into relative translational energy.

The attractive and repulsive character of surfaces   The question of the amount of the energy released in a chemical reaction that goes into internal (vibrational and rotational) energy is evidently linked very closely to the form of the potential-energy surface.   It has been seen (page 172) that for the sodium-halogen reactions a high percentage of the heat evolved goes into vibration of the bond formed, which was interpreted by Evans and M. Polanyi on the basis that the potential-energy surface is an *attractive* one.   On an attractive surface, an example of which is shown in Fig. 58, the heat of reaction is liberated along the coordinate corresponding to the approach of atom A to the BC molecule.   On such a surface, a typical reaction path is as shown schematically in Fig. 59, and it is to be seen that there is much vibrational energy in the AB molecule formed.

By contrast, on a purely repulsive surface the entire heat of reaction

[1] *Ibid.*, p. 376.

[2] J. K. Cashion and J. C. Polanyi, *Proc. Roy. Soc. (London)*, **A258**:570 (1960); P. E. Charters and J. C. Polanyi, *Discussions Faraday Soc.*, **33**:107 (1962); F. D. Findlay and J. C. Polanyi, *Can. J. Chem.*, **42**:2176 (1964); J. C. Airey, R. R. Getty, J. C. Polanyi, and D. R. Snelling, *J. Chem. Phys.*, **41**:3255 (1964); K. G. Anlauf, P. J. Kuntz, D. H. Maylotte, P. D. Pacey, and J. C. Polanyi, *Discussions Faraday Soc.*, **44**:183 (1967).

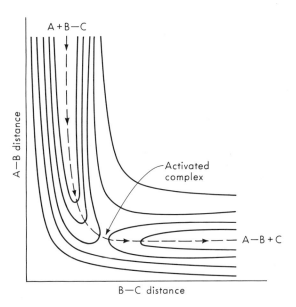

Fig. 63 A repulsive surface, showing a typical reaction path. Little energy now goes into vibration of AB; most goes into relative translational energy.

is liberated along the coordinate which corresponds to increasing separation of the products AB + C. Such a surface is shown schematically in Fig. 63; a typical reaction path is indicated, and there is seen to be no vibrational energy in the product AB, all of the heat of reaction going into relative translational energy.

These two cases, of purely attractive (Fig. 58) and purely repulsive (Fig. 63) surfaces, of course represent extremes, which are probably never realized; actual potential-energy surfaces will lie between these extremes and will have more or less attractive character. Apart from this, even on purely repulsive surfaces not all reactions would follow the rectilinear path represented in Fig. 63, and inevitably some of the heat of reaction would on the average pass into vibrational modes. In spite of these factors, the concept of attractive and repulsive surfaces has proved useful in relating the form of surfaces to the results of molecular-beam and chemiluminescence experiments.

This aspect of the problem has been considered in particular by Blais and Bunker and by Polanyi and coworkers. Blais and Bunker[1] considered moderately attractive and moderately repulsive potential-energy surfaces for the reaction $K + CH_3I$ and calculated a large number of trajectories with CH_3 considered a single particle (compare the calcula-

[1] N. C. Blais and D. L. Bunker, *J. Chem. Phys.*, **37**:2713 (1962); **39**:315 (1963).

tions of Karplus and Raff, to be discussed on pages 194 to 197). They found that the experimental result that the energy is released as internal energy is obtained only when the moderately attractive surface was used. The earlier calculations related to collisions in a plane, but Bunker and Blais[1] observed that the results were the same when motion in three dimensions was permitted. These workers have also given special consideration to how the masses of the species A, B, and C will affect the distribution of energy in the products of reaction. They made calculations for hypothetical reactions in which the masses of species A, B, and C could be light, medium, or heavy; light corresponded to 2 atomic-mass units; medium, to 16; and heavy, to 128. Most of their calculations were made with respect to a moderately attractive surface, and they found that almost all mass combinations led to much of the energy's passing into vibration. These calculations therefore support the conclusion that the results for the $H + X_2$ reactions, in which little energy goes into vibration, require that the surface be repulsive.

J. C. Polanyi and coworkers[2] made similar calculations based on the series of potential-energy surfaces shown in Fig. 64. These vary from a largely attractive surface (a) to a largely repulsive one, and the calculations were for collisions occurring in one plane (2D). These calculations indicate that even in this 2D case, with bent activated complexes allowed, there is still a useful correlation between the degree of attractive character of a surface and the amount of energy passing into vibration, provided that the attacking atom A is light compared with B and C (for example, $H + X_2$). In such systems, for example, if the surface is highly repulsive, little energy passes into vibration, as predicted by the simplest considerations (see Fig. 63).

When, however, the attacking atom A is not light compared with B and C the situation is more complicated; the calculations now indicate that even on a very repulsive surface a considerable amount of energy can still pass into vibration. The difference between the situations is that if the attacking atom A is light it approaches B rapidly so that the normal A—B bonding distance has been reached before there is much release of B-C repulsion. Consequently, when the B-C repulsion energy is released it tends to produce recoil of AB as a whole, which means that most energy passes into translation. When, on the other hand, A is heavy it approaches BC more slowly, so that there is time for the B-C repulsion to be released while the A—B bond is still extended (in terms of Fig. 63

[1] D. L. Bunker and N. C. Blais, *J. Chem. Phys.*, **41**:2377 (1964).

[2] J. C. Polanyi and S. D. Rosner, *J. Chem. Phys.*, **38**:1028 (1963); J. C. Polanyi, *J. Quant. Spect. Radiative Transfer*, **3**:471 (1963); P. J. Kuntz, E. M. Nemeth, J. C. Polanyi, S. D. Rosner, and C. E. Young, *J. Chem. Phys.*, **44**:1168 (1966).

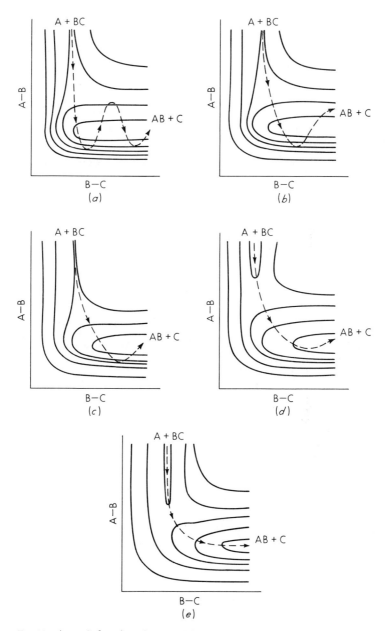

Fig. 64 A graded series of potential-energy surfaces, (a) attractive and (e) repulsive, the others being intermediate in character. The dashed lines show, *purely schematically*, a typical reaction path.

this means that the reaction is not following the path shown, but is "cutting the corner"). This release of repulsion energy while the A—B distance is still large gives rise to vibrational excitation of A—B, since the effect of the repulsion-energy release is to push B toward A. This situation, termed *mixed* energy release, is illustrated schematically in Fig. 65.

It is evident that some very important conclusions arise from considering the degree of the attractive or repulsive character of a potential-energy surface. This classification of surfaces, however, represents only a very first step toward a more detailed understanding of the features of chemical reactions.

MOLECULAR BEAMS[1]

The essential feature of a molecular-beam experiment is that collimated beams of atoms or molecules are caused to cross one another. Movable detectors determine the directions taken by the product molecules and by

[1] For reviews see D. R. Herschbach, *Appl. Opt., Suppl. 2, Chemical Lasers,* **1965**:128; *Advan. Chem. Phys.,* **10**:319 (1966); E. F. Greene, A. L. Moursund, and J. Ross, *Advan. Chem. Phys.* **10**:135 (1966); A. R. Blythe, M. A. D. Fluendy, and K. P. Lawley, *Quart. Rev. (London),* **20**:465 (1966); J. P. Toennies, in H. Hartmann and J. Heidberg (eds.), "Chemische Elementarprocesse," p. 157, Springer Verlag, Berlin, 1968; R. B. Bernstein, *Science,* **144**:141 (1964); E. F. Greene and J. Ross, *Science,* **159**:587 (1968); B. H. Mahan, *Accounts Chem. Res.,* **1**:217 (1968).

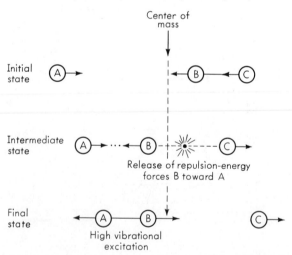

Fig. 65 A schematic representation showing how, even on a repulsive surface, if atom A is not relatively light, the release of repulsion energy can produce considerable vibrational excitation in the product AB (*mixed* energy release).

the unreacted molecules. Analysis of the experimental results yields information about the distribution of energy and angular momentum among the reaction products, the dependence of the total cross section on the molecular energies, the quantum states taken up by the products, the lifetime of the collision complex, and a number of other important features. In an ideal molecular-beam experiment the measurements would be performed on monoenergetic beams of molecules in selected rotational and vibrational states, and they would be repeated for a number of quantum states of the reactants. In practice, however, the intensity of the incident beams would be too greatly reduced for such detailed energy selection, and there are also difficulties in analyzing a beam into vibrational states. As a result, it has proved necessary to work with less selected beams and carry out a more detailed analysis of the scattering.

The type of scattering that occurs in a molecular-beam experiment may conveniently be discussed with reference to a reaction of the type

$$M + RX \rightarrow MX + R$$

Suppose that M is initially in the ith quantum state, and RX in the jth; the following three types of scattering may then occur:

Elastic

1. $M(i) + RX(j) \rightarrow M(i) + RX(j)$

Inelastic

2. $M(i) + RX(j) \rightarrow M(k) + RX(l)$

Reactive

3. $M(i) + RX(j) \rightarrow MX(m) + R(n)$

In elastic scattering (Eq. 1) there is no change in the quantum states of the colliding molecules or in the speed of their relative motion; the direction of their relative motion may, however, change. In inelastic scattering (Eq. 2) there is no change of chemical species, but the reactants emerge from the collision in quantum states which are different from the original ones. In reactive collisions (Eq. 3) chemical change has occured. It is possible, in a molecular-beam experiment, to study all three kinds of scattering, each one of which provides information about how the molecules interact.

The information that is provided by the results of molecular-beam experiments may be classified as follows:

1. The probability that M and RX interact at all as they pass each other. This can be expressed in terms of the cross section for a particular process, and the experiment will provide information about the

variation of the cross section with the relative energy of the molecules.

2. The probability of scattering to a particular solid angle, and the way in which this probability varies with the energy.
3. The speeds with which the products separate after reaction.
4. The internal states of the products formed in the reaction.
5. The threshold energy for reaction.
6. The way in which the inelastic and reactive scattering depend on the initial relative orientation of the colliding molecules.
7. The lifetime of the collision complex formed.
8. The relationship between the scattering and the electronic and geometric structures of the reactants.

Figure 66 shows in a very schematic way the kind of apparatus used in a molecular-beam experiment. With the movable detector the amount of scattered M or MX (for example, K or KI) can be determined as a function of the scattering angle α.

The first investigation of the kinetics of a chemical reaction by the molecular-beam technique was described in 1954 by Bull and Moon,[1] who studied the Cs + CCl$_4$ systems. Shortly afterward Taylor and Datz[2]

[1] T. H. Bull and P. B. Moon, *Discussions Faraday Soc.*, **17**:54 (1954).

[2] E. H. Taylor and S. Datz, *J. Chem. Phys.*, **23**:1711 (1955); see S. Datz and E. H. Taylor, in I. Estermann (ed.), "Recent Advances in Molecular Beams," p. 157, Academic Press, Inc., New York, 1959.

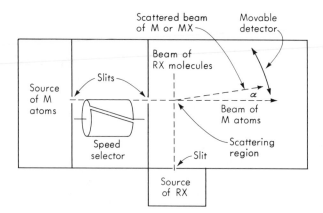

Fig. 66 Schematic diagram of molecular-beam apparatus for the study of a reaction of the type M + RX → MX + R. The angle α is the so-called "laboratory" scattering angle.

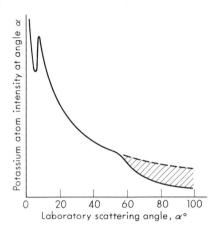

Fig. 67 Typical scattering pattern for reactions of the type K + HX or K + CH₃X. The shaded area indicates the loss of K due to chemical reaction.

studied the reaction

$$K + HBr \rightarrow KBr + H$$

and investigated the angular distribution of the products of reaction. They found the collision yield, i.e., the ratio of KBr molecules detected to K atoms scattered, to be about 10^{-3}. From the variation of this yield with beam "temperatures," calculated from the energy and other characteristics of the beam, Taylor and Datz estimated the energy of activation of the reaction to be about 3 kcal/mole. In a later study of the same reaction Greene, Roberts, and Ross[1] used more refined apparatus and obtained more detailed information about the collision yields.

Since these early investigations a variety of reactions have been studied in molecular beams, particularly by R. B. Bernstein, S. Datz, E. F. Greene, D. R. Herschbach, J. Ross, and coworkers. A great deal has been learned from these studies about the detailed mechanisms of these reactions. It is outside the scope of the present book to give a complete review of this work, and only a few examples can be mentioned. Herschbach and his coworkers[2] and Greene and Ross[3] have studied the scattering of K atoms from a variety of halogen-containing compounds such as the hydrogen halides, methyl iodide, carbon tetrachloride, and bromine. In some cases (K + CH₃I, K + HBr, K + CCl₄) the scattering pattern was of the type shown in Fig. 67. The marked loss of intensity at high angles of scattering is attributed to the removal of K by reaction.

[1] E. F. Greene, R. W. Roberts, and J. Ross, *J. Chem. Phys.*, **32**:940 (1960).

[2] For reviews see D. R. Herschbach, *Appl. Opt., Suppl. 2, Chemical Lasers*, **1965**:128; in J. Ross (ed.), "Molecular Beams," p. 319, Interscience Publishers, Inc., New York, 1966.

[3] See E. F. Greene and J. Ross, *Science*, **159**:587 (1968).

It is assumed that the elastic scattering at small angles, which arises from collisions in which the reactant molecules do not come close to one another, is unaffected by the reaction process, and on this basis the reaction parameters can be used to predict the complete elastic-scattering pattern in the absence of reaction.   The total number of K atoms lost per second can then be obtained from the difference between the predicted and observed signals.   The absolute total-reaction cross section is given by the number of atoms lost per second divided by the incident flux of target molecules and the volume of intersection of the beams.

In their investigations of the reaction

$$K + HBr \rightarrow KBr + H$$

Greene and Ross and their coworkers have studied, as a function of scattering angle, both the scattered K and the KBr.   The reaction cross section obtained from the K atom scattering was 34 Å^2, and the activation energy was 1 kcal/mole.   A somewhat lower value of the cross section was provided by the results on the KBr formation.   For the analogous reaction

$$K + HI \rightarrow KI + H$$

the cross sections obtained by the two methods were 31 Å^2 and 14 Å^2 respectively.

The reaction cross sections obtained, by using incident energies of about 1 kcal/mole, range from about 10 Å^2 for some of the alkyl halides (e.g., for K + CH_3I) to over 100 Å^2 for Cs + Br_2, K + Br_2, Cs + ICl, Cs + IBr, and Cs + I_2.   The results from a number of different systems fall largely into two groups, those with large ($\gtrsim 100$ Å^2) and those with small ($\lesssim 10 \text{Å}^2$) reaction cross sections.   The same division into two classes is revealed by an examination of the angular distribution of the scattered product.   This is illustrated in Fig. 68 for the K + CH_3I and K + Br_2 systems.   In K + Br_2 and all processes of the first type, where the cross sections are large, the alkali halide produced continues to travel in much the same direction as the beam of K atoms.   This behavior is represented in Fig. 69.   In the reactions of small cross sections, on the other hand, the alkali halide molecules are scattered backward from the center of mass; in other words, an observer traveling on the center of mass would see the two reactants approaching from his left and right and colliding at his point of observation; he would then observe the alkali halide returning roughly in the direction from which the alkali atom came.   This behavior is illustrated schematically in Fig. 70.

Only a small number of experiments have involved measuring the velocity distribution in one of the product species; this is a direct way of obtaining information about the partitioning of energy between internal

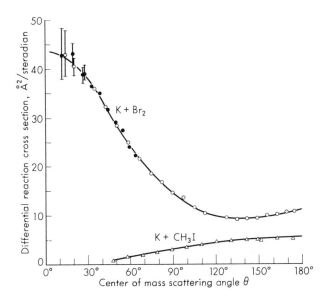

Fig. 68 Angular distributions, in the center-of-mass systems, of the alkali halide product from the K + CH₃I and K + Br₂ reactions. Note that the angle θ has a frame of reference moving with the center of mass, as opposed to α in Fig. 67 which is the directly observed or *laboratory* angle of scattering.

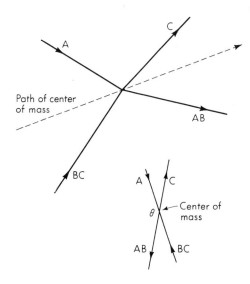

Fig. 69 Directions of motion of reactants and products, for a stripping mechanism. The upper diagram shows actual directions; the lower, the directions relative to the center of mass.

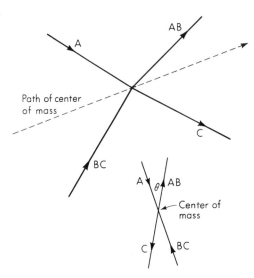

Fig. 70 Directions of motion of reactants and products, for a rebound mechanism. The upper diagram shows actual directions; the lower, the directions relative to the center of mass.

and translational modes. Such measurements have been made for KBr scattered from crossed beams of velocity-selected K atoms and thermal HBr and Br_2 beams. At any angle of observation the KBr velocity was found to be distributed about a well-defined maximum, the most probable velocity varying somewhat with the angle of observation.

Steric effects   The first work on the reaction

$$K + CH_3I \rightarrow KI + CH_3$$

was carried out by Herschbach, Kwei, and Norris,[1] who studied the angular distribution of the KI formed. More recently Brooks and Jones[2] and Beuhler, Bernstein, and Kramer[3] have refined these experiments by using beams of CH_3I molecules which were oriented by a powerful electric field; in this way the K atoms could be made to approach the CH_3I molecules principally from the I end or principally from the CH_3 end. The results show, as expected, that the ratio of the reaction cross sections for approach to the I and CH_3 ends is approximately 1.5; in other words, there is a definite *steric effect*. A dynamical calculation for this situation has been made by Karplus and Godfrey.[4]

[1] D. R. Herschbach, G. H. Kwei, and J. A. Norris, *J. Chem. Phys.*, **34**:1842 (1961).

[2] P. R. Brooks and E. M. Jones, *J. Chem. Phys.*, **45**:3449 (1966).

[3] R. J. Beuhler, R. B. Bernstein, and K. H. Kramer, *J. Am. Chem. Soc.*, **88**:5331 (1966); cf. K. H. Kramer and R. B. Bernstein, *J. Chem. Phys.*, **42**:767 (1965).

[4] M. Karplus and M. Godfrey, *J. Am. Chem. Soc.*, **88**:5332 (1966).

Stripping and rebound mechanisms Most of the reactions that have been studied by the molecular-beam method fall into one of two main classes:

1. Reactions of high cross sections ($\gtrsim 100$ Å²), in which the molecular product is scattered forward with respect to the center of gravity, i.e., *stripping* reactions (Fig. 69)
2. Reactions of low cross sections ($\lesssim 10$ Å²), in which there is backward scattering of the molecular product with respect to the center of gravity, i.e., *rebound* reactions (Fig. 70)

A stripping reaction of a rather extreme type is referred to as a *spectator-stripping* reaction. This term is borrowed from the theory of certain nuclear reactions.[1] It was first applied to a chemical process by Henglein and his coworkers;[2] their work was concerned with certain ion-molecule reactions, and they concluded that a spectator-stripping process occurs when the relative energy of the reactants is high. The characteristic of a spectator-stripping reaction A + BC → AB + C is that there is very little impulsive interaction between A and the remote atom C; the collision of A with BC thus has little effect on C, which therefore plays the role of a "spectator" to the interaction between A and B. The products separate so rapidly that there is no time for much of the momentum of the impact between A and B to be transferred to C. Once the interaction has set in, A + B and C behave as separate dynamic systems.

When spectator stripping occurs, an observer riding on the center of mass of the A-B-C system will therefore first see A and BC approaching him; A strips off B, and C continues undisturbed in its original direction, i.e., in the opposite direction to the approach of A. This is illustrated in Fig. 69. The spectator atom C is said to be *backward*-scattered, and AB to recoil *forward*, with reference to the center of mass.

We have seen that an important feature of a spectator-stripping mechanism is that the products must separate so rapidly that there is not enough time for C to be affected. If follows that spectator-stripping mechanisms will be favored if the reactants come together with very high energies. At these energies the time scale is much shorter than for

[1] See S. T. Butler, "Nuclear Stripping Reactions," John Wiley & Sons, Inc., New York, 1957; C. A. Levinson and M. K. Banerjee, in F. Ajzenberg-Selove (ed.), "Nuclear Spectroscopy," Chap. 5B, Academic Press, Inc., New York, 1960.

[2] A. Henglein and G. A. Muccini, *Z. Naturforsch.*, **17a**:452 (1962); **18a**:753 (1963); A. Henglein, K. Lacmann, and G. Jacobs, *Ber. Bunsenges. Physik. Chem.*, **69**:279, (1965); K. Lakmann and A. Henglein, *Ber. Bunsenges. Physik. Chem.*, **69**:286, 291 (1965); A. Henglein, K. Lacmann, and B. Knoll, *J. Chem. Phys.*, **43**:1048 (1965); for a general review see A. Henglein, *Advan. Chem. Ser.*, **58**:63 (1966).

momentum transfer within BC.   There is much qualitative evidence for stripping in molecular-beam experiments,[1] but for ordinary thermal reactions the abrupt transition postulated in the spectator model would seem to be out of the question, and even in cases which involve the harpooning mechanism (see page 172) the interaction time is not so short as to preclude significant momentum transfer to atom C.   The question of the relationship between the occurrence of a spectator-stripping reaction and the energy of impact is considered in a later section (page 197).

Collision times   It has been seen (page 169) that reaction mechanisms can be interpreted by two limiting models: the long-lived collision-complex or compound-state model, in which the lifetime of the complex is long compared with the rotational and vibrational periods, and the direct-interaction model, in which the collision time is approximately the time required for the reactant molecule to pass one another if there were no interaction.   Reactions of the first type may be said to be *indirect;* those of the second type, *direct* or *impulsive.*   These are the extreme mechanisms, and intermediate situations are, of course, possible.   The dynamical calculations for the $H + H_2$ reaction (page 170) have shown that the mechanism is direct.

Most of the reactions studied in molecular beams show a pronounced forward-backward asymmetry as far as the angular distribution of products is concerned.   The extreme cases of this were shown in Figs. 69 and 70 for stripping and rebound mechanisms, respectively. This asymmetry shows that the products "remember" the direction of the initial velocities.   This would not be the case if the collision complexes were able to rotate through more than half a turn before they broke down.   Reactions which behave in this way thus proceed by direct or impulsive mechanisms, and the molecular-beam studies lead to an upper limit of $\sim 5 \times 10^{-13}$ sec for the average lifetime of a collision complex for many of the reactions studied.

Miller, Safron, and Herschbach[2] have, however, investigated a family of reactions, including

$$Cs + RbCl \rightarrow CsCl + Rb$$

and

$$CsCl + KI \rightarrow CsI + KCl$$

[1] J. H. Birely and D. R. Herschbach, *J. Chem. Phys.*, **44**:1690 (1966); T. T. Warnock, R. B. Bernstein, and A. G. Grosser, *J. Chem. Phys.*, **46**:1685 (1967); E. A. Entemann and D. R. Herschbach, *Discussions Faraday Soc.*, **44**:299 (1967).

[2] W. B. Miller, S. A. Safron, and D. R. Herschbach, *Discussions Faraday Soc.*, **44**:108, 292 (1967).

for which there is symmetry in the angular distribution of products. This result indicates that reaction occurs with a complex which, for collision energies of 1 to 2 kcal/mole, exists long enough to undergo at least several rotations. A calculation[1] of the potential-energy surface for the K-Na-Cl system indicates that it is highly attractive and there is a basin about 13.5 kcal/mole deep compared with the energy of the products. The dimerization energies of alkali halides have been determined from thermochemical studies, and a basin about 30 to 50 kcal deep is expected for the Cs-Cl-KI system. The existence of such basins would provide an explanation for these long-lived complexes. The evidence indicates very large cross sections, $\gtrsim 150$ Å^2, for these systems. It is estimated that the lifetimes of the complexes are at least 5×10^{-12} sec, that is, several times the rotational period of 1 to 2×10^{-12} sec. Examples in which the complex breaks up in about one rotational period have been found[2] for the Cs + TlX → CsX + Tl systems (X = Cl and I), where the lifetimes are $\sim 10^{-11}$ sec.

It is not yet clear whether the existence of a potential-energy basin is essential to the occurrence of these long-lived complexes.

Partitioning of energy   It has been seen (page 171) that the chemiluminescence studies lead to important conclusions about the distribution of energy among products; for the reaction

$$H + Cl_2 \rightarrow HCl + Cl$$

for example, a good deal of the energy released passes into translation and rotation of the products, the remainder passing into vibration.

The molecular-beam work also reveals information of this kind, although rather a detailed analysis is necessary. The earlier results on reactions such as

$$K + Br_2 \rightarrow KBr + Br$$
$$K + CH_3I \rightarrow KI + CH_3$$
$$Cs + CH_3I \rightarrow CsI + CH_3$$

were originally interpreted as indicating that much of the energy released went into vibration, with only a relatively small amount going into translation and rotation. These conclusions were based on qualitative dynamical analysis of angular-distribution measurements. More recently, direct velocity analysis of the products has been made for several reac-

[1] M. S. Child and A. Roach, *Mol. Phys.*, **14**:1 (1968); cf. M. S. Child, *Discussions Faraday Soc.* **44**:68 (1967).

[2] G. A. Fish, J. D. McDonald, and D. R. Herschbach, *Discussions Faraday Soc.*, **44**:228 (1967).

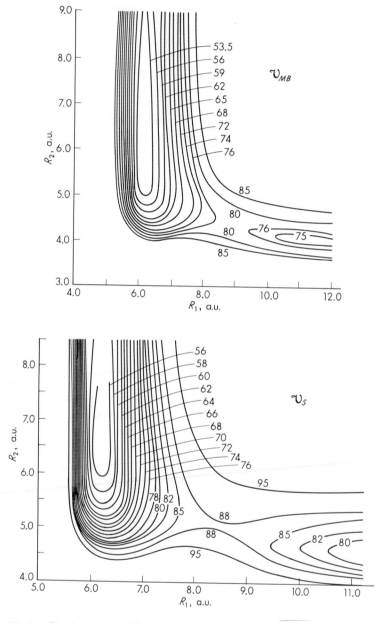

Fig. 71 The four potential-energy surfaces employed in the calculations by Karplus and Raff. R_1 is the K—I distance, and R_2, the I—CH₃ distance.

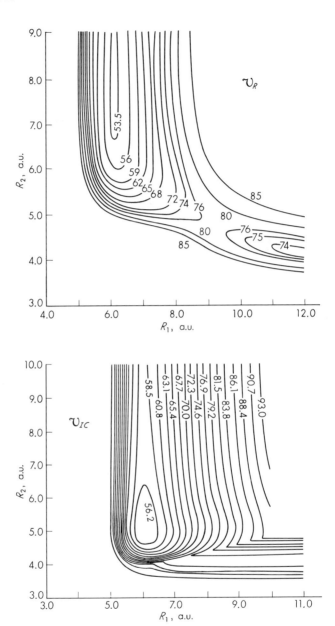

Fig. 71 (*Continued.*)

tions, including those listed above. Entemann and Herschbach[1] have discussed the interpretation of these results. The original interpretation is supported except in a few unfavorable cases for which the original analysis underestimated the amount of energy passing into translation. For the $K + Br_2$ reaction, over 90 percent of the exothermicity goes into internal excitation of the KBr; for $K + CH_3I$, about 70 percent goes into internal excitation (the partitioning between KI and CH_3 is not known); for $Cs + CH_3I$, only about 40 percent goes into internal excitation.

Dynamical calculations   In view of the molecular-beam studies made on reactions involving alkali-metal atoms, it is clearly of interest for dynamical calculations to be made for such reactions, with realistic potential-energy surfaces. Such calculations have been made by Blais and Bunker[2] and by Karplus and Raff.[3] Part of the object of this work was to gain some insight into the relationship between the assumed potential-energy surface and the characteristics of the reaction.

The calculations by Karplus and Raff were made for four potential-energy surfaces, which are shown in Fig. 71. The first, $\mathcal{U}_{MB}$, is expressed by the equation

$$\mathcal{U}_{MB} = D_1(1 - e^{-\beta_1'(r_1-r_{01})})^2 + D_2(1 - e^{-\beta_2(r_2-r_{02})})^2$$
$$+ D_2[1 - \tanh(ar_1 + c)]e^{-\beta_2(r_2-r_{02})} + D_3 e^{-\beta_3(r_3-r_{03})} \qquad (11)$$

The first and second terms are Morse-like potentials for KI and CH_3I, respectively; r_1 is the K—I distance; and r_2 is the H_3C—I distance. The third term is an attenuation factor that reduces the CH_3I attraction energy as the K atom approaches. The fourth term introduces an exponential repulsion between the K atom and the CH_3 group. This surface resembles one originally introduced by Blais and Bunker.

Surface $\mathcal{U}_R$ differs from $\mathcal{U}_{MB}$ only in the method used for the long-range interaction between K and CH_3I. Surface $\mathcal{U}_S$ has the form

$$\mathcal{U}_S = D_1(1 - e^{-\beta_1''(r_1-r_{01})})^2 + D_2(1 - e^{-\beta_2(r_2-r_{20})})^2$$
$$+ D_2[1 - \tanh(ar_1 + c)]e^{-\beta_2(r_2-r_{02})}$$
$$+ \frac{\varepsilon}{1 - 6\alpha^{-1}}\left\{\frac{6}{\alpha}\exp\left[\left(1 - \frac{r}{r_m}\right)\alpha\right] - \left(\frac{r_m}{r}\right)^6\right\} \qquad (12)$$

[1] E. A. Entemann and D. R. Herschbach, *Discussions Faraday Soc.*, **44**:289 (1967).

[2] H. C. Blais and D. L. Bunker, *J. Chem. Phys.*, **37**:2713 (1962); **39**:315 (1963).

[3] M. Karplus and L. M. Raff, *J. Chem. Phys.*, **41**:1267 (1964); L. M. Raff and M. Karplus, *J. Chem. Phys.*, **44**:1212 (1966). In a later paper in *J. Chem. Phys.*, **44**:1202 (1966), Raff has considered the dynamics of four-atom systems. His treatment is applied to $K + C_2H_5I$, with C_2H_5 treated as two atoms.

The first three terms are identical in form with $\mathcal{U}_{MB}$. The fourth term is an "exponential-six" potential; r is the distance from K to the center of mass of CH_3I, and the parameters ε, α, and r_m correspond to the long-range effective K-CH_3 interaction. Since the center of mass of CH_3I is taken as the origin in the exponential-six potential-energy term, $\mathcal{U}_S$ has considerably more spherical symmetry than either $\mathcal{U}_{MB}$ or $\mathcal{U}_R$.

Surface $\mathcal{U}_{1C}$ is one having a basin. It is obtained by a superposition of covalent and ionic terms. It is rather unrealistic and was introduced to establish the effect of a basin; as will be seen, the calculations indicate that it does not correspond to the truth. Since there is a basin, the collision complexes frequently undergo a number of vibrations before decomposing, and this is contrary to the molecular-beam evidence.

The total reaction cross sections calculated for surfaces $\mathcal{U}_{MB}$, $\mathcal{U}_R$, and $\mathcal{U}_S$ are 24.6, 36.1, and 13.4 $\mathring{A}^2$, respectively. These values are quite insensitive to the relative energies of approach, a result of there being essentially no activation energy. The experimental cross section is about 30 $\mathring{A}^2$, a good deal smaller than the calculated values. This deficiency could probably be remedied by varying the potential-energy parameters somewhat.

In considering the reaction characteristics it should be borne in mind that some artificiality has been introduced by assuming the CH_3 group to be a single particle; one result of this is that it is impossible to investigate the distribution of the heat of reaction among the internal vibrational modes or to assess the effect of structure of the R radical on reactions of the type $K + RI$.

One reaction attribute of interest is the reaction probability $P_r(V_R,v,J,b)$, which is the probability that reaction occurs when V_R (relative velocity), v (vibrational quantum number), J (rotational quantum number), and b (impact parameter) are fixed at particular values. One can consider how P_r varies with, for example, the impact parameter b, with the other parameters V_R, v, and J held at their most probable values. The results of such calculations for the three surfaces $\mathcal{U}_{MB}$, $\mathcal{U}_R$, and $\mathcal{U}_S$ are shown in Fig. 72. It is to be seen that the curves for $\mathcal{U}_{MB}$ and $\mathcal{U}_R$ are similar; there is a flat region at low b values, in which P_r is less than unity, and as b becomes large, P_r approaches zero. By contrast, the $\mathcal{U}_S$ curve shows no plateau and falls toward zero at smaller b values, but approaches unity as b approaches zero. The falling-off at high b values is due to the decrease in the $K\cdots CH_3I$ attraction as b increases, this decrease being more rapid for the $\mathcal{U}_S$ surface. The reason that P_r does not approach unity at low b values for the $\mathcal{U}_{MB}$ and $\mathcal{U}_R$ surfaces is that for those there is a steep, short-range repulsion between K and CH_3; this prevents reaction when the K atom approaches CH_3 rather than I. At small b values reaction will occur only if K approaches the I atom. Sur-

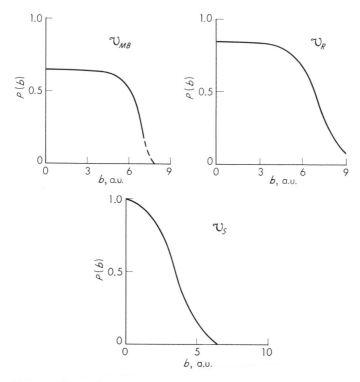

Fig. 72 The probability of reaction, P_r, as a function of impact parameter b, for the K + CH_3I reaction. (Calculations by Raff and Karplus.)

face $\mathcal{U}_S$ has a considerably weaker and symmetrical repulsion with its origin at the center of mass of the CH_3I molecule. As a result, reaction will still occur when K approaches the CH_3 group; in other words, there is no steric factor for this surface. The experimental evidence regarding the steric factor, from molecular-beam experiments, has been referred to on page 188.

The K + CH_3I reaction is exothermic by about 22 kcal/mole, which may appear either as internal excitation of the products or as kinetic energy of relative motion. Since CH_3 has been treated in the calculations as a single particle, the internal energy can only be present as rotational and vibrational energy in the KI molecules formed. The calculations with the three surfaces $\mathcal{U}_{MB}$, $\mathcal{U}_R$, and $\mathcal{U}_S$ indicate that the majority of the energy released appears as vibrational energy of the newly formed KI molecule. This result is easy to understand for the $\mathcal{U}_{MB}$ and $\mathcal{U}_S$ surfaces, both of which are of the *attractive* type (see Fig. 58); the energy is released during the approach of the reactants to one another. For the $\mathcal{V}_R$ surface, on the other hand, the situation is somewhat different, since along the path of minimum energy a significant fraction of the energy would be released after the system has turned the corner. The calculations do nevertheless indicate that much of the energy released passes into vibrational energy of the KI, even for this surface; this is of interest in view of the earlier discussion (page 182) of evidence that under certain circumstances one can get high vibrational excitation on a repulsive surface.

It has been seen (page 191) that the earlier work on this reaction did indicate that most of the energy released went into vibration, but that the more recent work shows that the division between vibration and translation is about equal. It would therefore appear that the three surfaces $\mathcal{V}_{MB}$, $\mathcal{V}_R$, and $\mathcal{V}_S$ do not have enough repulsive character to explain the results; further calculations are needed.

The molecular-beam experiments indicate (see page 186) that there is backward scattering of the product KI with respect to the center of mass of the system; i.e., the process occurs by a *rebound* mechanism. The dynamical calculations predict this result for the $\mathcal{V}_{MB}$, $\mathcal{V}_R$, and $\mathcal{V}_S$ surfaces. This result requires that the complex break up before it can rotate through a half-turn, i.e., the complex is of short life. This conclusion is confirmed by calculations on the details of the collisions. Figure 73 gives a typical trajectory based on surface $\mathcal{V}_{MB}$ and shows that the system passes directly through the activated state. The effect of introducing a basin in the potential-energy surface is shown for surface $\mathcal{V}_{IC}$ in Fig. 74, in which it is seen that the complex performs several vibrations before breaking down into $KI + CH_3$.

EFFECT OF COLLISION ENERGY

It has been seen in the previous sections of this chapter that bimolecular exchange reactions of the type $A + BC \rightarrow AB + C$ can proceed in a variety of ways; they may be direct (impulsive) or indirect, and they may involve stripping or rebound mechanisms. The factors determining the pattern of a given chemical reaction are by no means clear. Obviously the shape of the potential-energy surface has much to do with the detailed course of a reaction; the existence of a basin, for example, may cause the reaction to occur by a long-lived complex, while a highly attractive surface tends to favor a stripping mechanism.

Another factor, the collision energy, also influences the reaction pattern, as has been discussed in particular by J. C. Polanyi.[1] As the colli-

[1] J. C. Polanyi, *Discussions Faraday Soc.*, **44**:293 (1967).

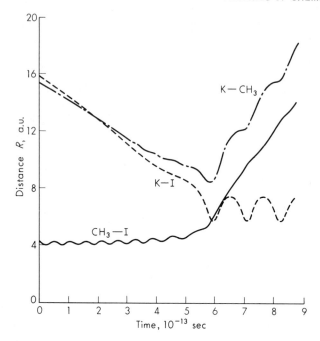

Fig. 73 A typical trajectory, for surface υ_{MB} of the K + CH_3I reaction.   (Calculations by Raff and Karplus.)

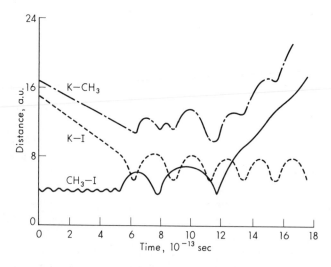

Fig. 74 A typical trajectory for the surface υ_{1C}, which has a basin.   (Calculations by Raff and Karplus for the K + CH_3I reaction.)

sion energy increases, any reaction will tend to approach the spectator-stripping type of behavior. It has been seen that in spectator stripping there is little impulsive interaction between A and the remote atom C, the products AB and C separating so rapidly that there is no time for much of the momentum of the impact between A and B to be transferred to C. This situation is favored if the collision energy is large. It must be realized, however, that not all collisions of high energy will lead to stripping; for example, if A strikes B along the B-C axis (i.e., if the impact parameter b is zero), there cannot be pure spectator stripping, since this would require C to pass right through B for its motion to continue undisturbed; instead, there will be backward scattering. However, although there can never be pure spectator stripping for every kind of collision even at high energies, spectator stripping will on the whole become more important as the collision energy is increased. Polanyi suggests, in order to take account of the inevitability of a certain amount of backward scattering, that the term spectator stripping be applied to systems in which more than 90 percent of the reactive collisions lead to scattering in the forward hemisphere. This leads to a definition of the spectator-stripping threshold energy, which is the collision energy at which this 90 percent criterion becomes satisfied.

A spectator-stripping process is necessarily *direct*, by which is meant one in which the old bond B—C is continuously extended as the reaction proceeds. It seems likely that a reaction cannot occur via a long-lived complex unless there is a basin in the potential-energy surface, but even if there is a basin, high collision energies will tend to cause the system to pass right through the activated state and they will therefore favor spectator stripping. At low collision energies, reaction via a long-lived complex will be favored, although it will not be realized if the potential-energy surface is not of the right form. Reaction via a long-lived complex, i.e., an *indirect* reaction, has been seen to produce symmetrical angular scattering in the frame of reference of the center of mass. A reaction could therefore be described as indirect if it approximated symmetric angular scattering to a prescribed extent. The threshold energy for indirect reaction would then be defined as the collision energy below which this criterion was satisfied. It may well be that this limit can never be found for the majority of reactions. Apart from the probable necessity for an energy basin, for any reaction with an activation energy the reaction cross section may, as the collision energy is lowered, become vanishingly small before the threshold for indirect behavior is reached. For these reasons the threshold for indirect behavior is not as useful as that for stripping, which will now be considered in more detail.

The threshold energy for stripping provides useful information about the reaction dynamics. Stripping requires that the new molecule AB shall not interact with C; a low threshold energy for stripping therefore implies that any heat evolved is liberated as attraction between A and B, and not as repulsion between AB and C. For reactions *without an activation energy* this can be satisfied if:

1. No heat is evolved ($-\Delta H = 0$).
2. The potential-energy surface is of the attractive type, which means that the repulsive energy release is small.

For reactions *with an activation energy* it is necessary in addition to either (1) or (2) that:

3. The barrier should correspond to little B—C extension, as in the case shown in Fig. 58.

Normally we have some information on these points, and can therefore choose to some degree between (1), (2), and (3).

On the basis of the above discussion we may now consider six families of A + BC reactions, to many of which reference has already been made in this chapter. These families are arranged as far as possible (with a large element of guesswork) in the order of increasing threshold energy for stripping.

1. *Alkali-metal plus halogen-molecule reactions*   These systems have been considered earlier (page 172); they have been seen to occur by a harpooning mechanism, which favors spectator stripping. Important confirmation that a harpoon model applies has been provided by the molecular-beam studies of Warnock, Bernstein, and Grosser.[1] The energy threshold for stripping is clearly very low. Polanyi[2] has also pointed out that, since the greater part of the energy of reaction will have been liberated by the time A^+ and B^- have reached normal bonding distance, there is little room for energy change along the B-C coordinate of the potential-energy surface (see Fig. 62). This tends to favor *indirect* reaction, as well as stripping. The classical dynamical calculations in fact show that with a sufficiently small B-C interaction the trajectories for thermal collision energy show evidence of some indirect action, i.e., secondary encounters among the reacting atoms, as well as of stripping.

2. *Alkali-metal plus alkali halide reactions*   These reactions (see page 190) were first studied by Miller, Safron, and Herschbach[3] and shown to

[1] T. T. Warnock, R. B. Bernstein, and A. E. Grosser, *J. Chem. Phys.*, **46**:1685 (1967).

[2] J. C. Polanyi, *Discussions Faraday Soc.*, **44**:293 (1967).

[3] W. B. Miller, S. A. Safron, and D. R. Herschbach, *Discussions Faraday Soc.*, **44**:108 (1967).

proceed through a long-lived complex; at 1 to 2 kcal/mole collision energy, this complex survives several rotations. The energy threshold for indirect reaction is probably much higher than for alkali-metal plus halogen-molecule reactions. This can be understood from the low energy evolved in the metal plus alkali halide reactions, so that there is little repulsion between the products. Calculations by Child and Roach[1] have indicated that there is a small basin in the potential-energy surface, which will raise the energy threshold for complex reaction and for stripping. So far the experiments have only been done in the lower-energy range, where the reaction is indirect; it will be of interest to see whether, as predicted, an increase in the collision energy will lead to a direct, stripping mechanism.

3. *Ion-molecule reactions of the type* $A^+ + BC \rightarrow AB^+ + C$ A reaction of this type, namely,

$$Ar^+ + D_2 \rightarrow ArD^+ + D$$

has been studied in crossed molecular beams by Wolfgang and coworkers,[2] and dynamical calculations have been made by Kuntz and Polanyi.[3] Even at collision energies of 2.3 kcal/mole the reaction is direct, as is the reaction

$$N_2^+ + D_2 \rightarrow N_2D + D$$

studied by Wolfgang et al.[2] and by Mahan and coworkers.[4] The stripping threshold for the $Ar^+ + D_2$ reaction is between 25 and 50 kcal/mole. The energy liberated in the reaction is 25 to 30 kcal/mole, most of which will be liberated as attraction between the reactants as they approach under the influence of charge plus induced-dipole forces. It is concluded, contrary to previous reports, that no long-lived complex is involved, and this suggests that the B-C interaction is repulsive; in (2) above, a small attraction was seen to raise the threshold for complex reaction to greater than thermal energy. Presumably the withdrawal of charge from between the nuclei in D_2, as Ar^+ approaches, leaves the D atoms still at bonding separations but with diminished bonding; repulsion results which exceeds the ion plus induced-dipole attraction between the products. It appears that there is no true basin in the potential-energy surface for this reaction.

[1] M. S. Child and A. Roach, *Mol. Phys.* **14**:1 (1968).

[2] Z. Herman, J. Kerstetter, T. Rose, and R. Wolfgang, *Discussions Faraday Soc.*, **44**:123.

[3] P. J. Kuntz and J. C. Polanyi, *Discussions Faraday Soc.*, **44**:180 (1967).

[4] W. R. Gentry, E. A. Gislason, Y. T. Lee, B. H. Mahan, and C. Tsao, *Discussions Faraday Soc.*, **44**:137 (1967).

By contrast, a study by Henchman and coworkers[1] of the reactions

$$HD^+ + HD \begin{cases} \nearrow H_2D^+ + D \\ \searrow HD_2^+ + H \end{cases}$$

has revealed reaction through a long-lived complex at collision energies as high as 40 kcal/mole.

4. *Hydrogen-atom plus hydrogen-molecule reactions*   All the reactions in classes 1 to 3 above have large reaction cross sections, of about 100 Å^2. It has been seen earlier (page 189) that spectator stripping is favored if cross sections are large; the energy thresholds for stripping therefore tend to be lower if the potential-energy surface is one which allows reaction to occur at high separations between the reactants.

The reactions in the three classes 4, 5, and 6 have cross sections one or two orders of magnitude smaller than those in the first three.   In the case of reactions of the type

$$H + H_2 \rightarrow H_2 + H$$

this is due to the presence of an energy barrier, which results in a negative energy slope along the B-C coordinate.   The barrier is symmetrically placed and the descent from the col is accompanied by a release of B-C repulsion.   In a significant fraction of reactive encounters this repulsion is expected to cause the newly formed H_2 molecule to recoil backward along the direction from which the H atom approached.   The result of this is an increased threshold energy for stripping.

There are still no molecular-beam results which provide information about this threshold energy.   The classical and quantum-mechanical dynamical calculations by Karplus and coworkers (see pages 160 to 171) have shown that the molecular product is scattered backward (rebound) in the center-of-mass system at low collision energies and forward (stripping) at high collision energies.   The energy threshold corresponding to 90 percent scattering into the forward hemisphere was 1.3 to 1.4 ev.   The barrier height is about 9 kcal/mole, so that this is the repulsive energy release along the B-C coordinate of the potential-energy surface.

5. *Halogen-atom plus hydrogen halide reactions*   Reactions of the type

$$X + HY \rightarrow XH + Y$$

for example,

$$Cl + HI \rightarrow HCl + I$$

[1] L. Matus, I. Opansky, D. Hyatt, A. J. Masson, K. Birkenshaw, and M. J. Henchman, *Discussions Faraday Soc.*, **44**:146 (1967).

have been studied by spectroscopy.[1] The results have shown that practically all the energy released resides in vibrational and rotational excitation of the product molecule, so that relative translation between HCl and I must be slight.

At first sight this result suggests that the potential-energy surface is not of the repulsive type. The matter is, however, not as simple as this. The attacking atom in this system is heavy, and in these circumstances, for a more or less collinear collision, much of the B-C repulsion will be released while atoms A and B are still at an extended separation. This type of energy release, referred to as *mixed energy release*,[2] favors internal excitation in AB. Thus, provided the potential-energy surface favors reaction by collinear collisions, the observation of high vibrational excitation in the new bond does not exclude repulsive energy release.

There is no molecular-beam work on these reactions, but the following would be expected if indeed these reactions proceed on a repulsive surface. In view of the small reaction cross section, there will be mainly backward scattering (rebound) at thermal collision energies. At the higher collision energies required for stripping, the mixed energy release will be relatively less effective since the B—C distance will increase after A has come close to BC. The repulsive slope along the B-C coordinate will therefore raise the energy threshold for stripping.

6. *Hydrogen-atom plus halogen-molecule reactions*   Reactions of the type

$$H + X_2 \rightarrow HX + X$$

for example,

$$H + Cl_2 \rightarrow HCl + Cl$$

have also been studied spectroscopically.[3] In this class of reactions much less energy passes into internal energy (vibration and rotation) of the HX product. Two factors may contribute to this result:

1. The potential-energy surface may be of the repulsive type.
2. The attacking atom H is light and approaches too rapidly to allow a substantial amount of mixed energy release to occur.

It seems possible that the second factor above is operative. If the first is also important, we should expect an increase in the energy threshold for

[1] K. G. Anlauf, P. J. Kuntz, D. H. Maylotte, P. D. Pacey, and J. C. Polanyi, *Discussions Faraday Soc.*, **44**:183 (1967).

[2] J. C. Polanyi, *Appl. Opt.*, *Suppl.* 2, *Chemical Lasers*, **1965**:109.

[3] K. G. Anlauf, P. J. Kuntz, D. H. Maylotte, P. D. Pacey, and J. C. Polanyi, *Discussions Faraday Soc.*, **44**:183 (1967).

stripping, compared with the previous family of reactions. Preliminary molecular-beam experiments have been carried out, but the results are as yet inconclusive.

MORE COMPLEX REACTION

The reactions considered in the last section were of the three-center type $A + BC \rightarrow AB + C$. The dynamical problems are obviously more difficult if four or more centers are involved, as in reactions of the types

$$A + BCD \rightarrow AB + CD$$

and

$$A + BCD \rightarrow ABC + D$$

Smith[1] has made a study of a reaction of the first type, i.e.,

$$O + SCS \rightarrow OS + CS$$

and has measured the vibrational excitation in both products. About 8 percent of the energy released was found to pass into vibration in CS, and about 18 percent into vibration in SO. Smith has also carried out a dynamical study on various potential-energy surfaces and concludes that excitation of the CS results from release of energy as repulsion between the products. The yield is sharply dependent on the steepness of the interaction potential between the separating products and seems to vary with the impact parameters of the reactive collisions.

A spectroscopic investigation on a reaction of the second type has been made by Clough and Thrush[2] on the reaction

$$N + NO_2 \rightarrow N_2O + O$$

The bond-stretching modes in the product N_2O are found to be excited. However, although the N—N—O bond angle probably changes during separation of the products, it was found that there is very little excitation of the bonding mode. The distribution of energy between the two stretching modes suggests that the surface is more of the repulsive type.

The reaction

$$K + ICH_3 \rightarrow K + CH_3$$

has been discussed earlier in some detail (pages 194 to 197) and was seen to occur on a more or less repulsive surface. The dynamical calculations for this system have treated CH_3 as a single particle and therefore relate to a three-center reaction.

[1] I. W. M. Smith, *Discussions Faraday Soc.*, **44**:194 (1967).

[2] P. N. Clough and B. A. Thrush, *Discussions Faraday Soc.*, **44**:205 (1967).

So far very few dynamical calculations have been performed on four-center systems, and it is to be expected that this field will develop rapidly in the future.

UNIMOLECULAR DECOMPOSITIONS AND ASSOCIATION REACTIONS

As was seen in Chap. 6, most of the theories of unimolecular reactions are of the statistical type and not concerned with the detailed molecular dynamics. Part of the difficulty is that most unimolecular reactions for which data are available involve sufficiently complicated molecules that trajectory calculations are very difficult.

Bunker[1] has carried out a large number of Monte Carlo calculations for the decompositions of triatomic molecules, some of the calculations having special reference to the dissociations of N_2O and O_3. The main conclusions of his studies can be summarized as follows:

1. A model based on strict normal modes, with no anharmonicity and no flow of energy between modes, leads to unsatisfactory results, particularly in the low-pressure fall-off regions (see also page 145).
2. Theories of unimolecular decompositions based on the harmonic vibrations of the activated state, such as those of Giddings and Eyring[2] and Marcus,[3] lead to an overestimate of rate constants at a given energy by a factor of about 4. They also, however, underestimate the statistical factor of a given energy by a factor of about 2.5, so that the errors partly compensate for one another. On the whole the dynamical calculations support the RRKM theory (see page 122) within a factor of 2.
3. The calculations indicate that the vibrations of molecules of any size are subject to complete energy exchange in about 10^{-11} sec.
4. For highly energized molecules, less than half of the energy released appears, on the average, as product vibration.

Association reactions constitute a valuable source of vibrationally excited species which can be used in studies of various kinds. The vibrational relaxation, for example, can be studied, as also can the decomposition or isomerization of the excited species produced. Some studies have been made on the association of ground-state atoms to form electronically excited diatomic molecules:

$$A + B \rightarrow AB^*$$

[1] D. L. Bunker, *J. Chem. Phys.*, **37**:393 (1962); **40**:1946 (1964).
[2] J. C. Giddings and H. Eyring, *J. Chem. Phys.*, **22**:538 (1954).
[3] R. A. Marcus, *J. Chem. Phys.*, **20**:359 (1952).

Such a process is the inverse of a predissociation. Schiff and coworkers[1] have studied the reaction

$$H + O \rightarrow OH^*$$

and from the initial vibrational and rotational excitation in OH^* have drawn conclusions about the potential-energy curves. A theoretical treatment of radiative associations has been given by Palmer,[2] in terms of potential-energy curves and based on equilibrium statistical mechanics.

A molecular-beam study has been made by Fite and coworkers[3] of the reaction between sodium atoms and nitrogen molecules. The results are consistent with, but do not prove, the hypothesis that an association complex is first formed,

$$Na + N_2 \rightarrow NaN_2^*$$

and that this dissociates into $Na^* + N_2$. There is found to be a very efficient transfer of vibrational and rotational energy of the N_2 into electronic energy in the sodium. This reaction still awaits further theoretical study. An experimental investigation of the processes

$$Na^* + CO \rightarrow Na + CO^*$$
$$Na^* + NO \rightarrow Na + NO^*$$

has been made by Polanyi and coworkers,[4] who have measured infrared emission from the carbon monoxide molecules produced. Again, the results are to be explained in terms of association followed by dissociation.

QUANTUM-MECHANICAL TREATMENTS

The theories outlined so far in this chapter are largely classical ones; quantum restrictions are often introduced as initial conditions, but the dynamical calculations are based essentially on classical mechanics.

The present section is concerned with the attempts that have been made to devise purely quantum-mechanical theories and to apply them to experimental results. Unfortunately, solution of the wave equation is extremely difficult even for the simplest of the realistic potential-energy surfaces that have been developed. As a result, many of the quantum-mechanical treatments use rather simplified potential-energy surfaces.

[1] S. Ticklin, G. Spindler, and H. I. Schiff, *Discussions Faraday Soc.*, **44**:218 (1967).

[2] H. B. Palmer and R. A. Carabetta, *J. Chem. Phys.*, **46**:1538 (1967); H. B. Palmer, *J. Chem. Phys.*, **47**:2116 (1967).

[3] J. E. Mentall, H. F. Krause, and W. L. Fite, *Discussions Faraday Soc.*, **44**:157 (1967).

[4] G. Karl, P. Kruus, and J. C. Polanyi, *J. Chem. Phys.*, **46**:224 (1967); G. Karl, P. Kruus, J. C. Polanyi, and I. W. M. Smith, *J. Chem. Phys.*, **46**:244 (1967); J. C. Hassler and J. C. Polanyi, *Discussions Faraday Soc.*, **44**:182 (1967).

One of the earliest purely quantum-mechanical studies was that of Hirschfelder and Wigner,[1] who were concerned with the general problem of the probability of transmission across a potential-energy barrier. Golden and coworkers[2] applied a treatment to a simplified LEP surface for the $Br + H_2$ exchange reaction. The simplification consisted in assuming that the coulombic and exchange integrals between the outermost atoms Q and J are constant at some average values $\bar{Q}$ and $\bar{J}$. This is a significant modification of the LEP method, according to which the increase in J with the approach of A to BC makes an important contribution to the energy barrier to the reaction. In the present approximation the barrier will be largely due to the endothermicity of the reaction. Uncertainty in the choice of $\bar{Q}$ and $\bar{J}$ was found not to affect the rate by a factor of more than 2. The rate was more sensitive to the choice of the fraction of coulombic binding ρ, which was chosen to give the best fit to the experimental rate. Golden and Peiser summed the calculated quantum-mechanical transition probabilities between all possible initial and final quantum states and in this way obtained the rate constant as a function of the temperature; the calculated rates obeyed the Arrhenius law. These calculations led to the conclusion that 95 percent of the contribution to the rate came from hydrogen molecules in the first excited vibrational state. The newly formed HBr was found to have an effective rotational "temperature" of about one-half the initial temperature.

A somewhat similar quantum-mechanical study was made by Bauer and Wu,[3] who considered the reactions

$$Cl + H_2 \rightarrow HCl + H$$
and
$$Br + H_2 \rightarrow HBr + H$$

In their model they regarded the rate constant as a rate of activation, $k^{\ddagger}$, which involved the transformation of translational kinetic energy into vibrational energy, multiplied by the probability of passing from the activated state into the product state; this probability they took to be one-half (i.e., they were regarding the activated complexes as having an equal probability of passing in the two directions). They treated only collinear collisions and assumed that the interaction between A and BC was restricted to small distances. In place of a potential-energy surface they assumed an intermolecular potential and an intramolecular potential

[1] J. O. Hirschfelder and E. Wigner, *J. Chem. Phys.*, **7**:616 (1939).

[2] S. Golden, *J. Chem. Phys.*, **17**:620 (1949); S. Golden and A. M. Peiser, *J. Chem. Phys.*, **17**:630 (1949); *J. Phys. & Colloid Chem.*, **55**:789 (1951).

[3] E. Bauer and T. Y. Wu, *J. Chem. Phys.*, **21**:726 (1953); E. Bauer, Quantum Theory of Chemical Reaction Rates, *N.Y. Univ. Res. Rept.* CX-33 (1958).

and proceeded to calculate the translational-vibrational transition probability for passage of A + BC into the activated state. Their calculations led for both reactions to steric factors of about 8×10^{-3} (compared with the simple collision theory predictions). The lifetime of the activated state was calculated to be $\sim 5 \times 10^{-12}$ sec, which is long enough for 10 to 100 vibrations to occur. As we have seen, much shorter lifetimes are predicted by the classical calculations and the molecular-beam studies; other quantum-mechanical calculations (see below) also lead to very short lives. Mazur and Rubin[1] have also made calculations based on a highly idealized potential-energy surface.

More recently several workers have developed treatments of transmission probabilities over more realistic surfaces. Mortensen and Pitzer[2] employed a surface of the LEPS type and considered collinear collisions. Their calculations related to a variety of initial vibrational and translational energies. For collision energies considerably in excess of the threshold required for reaction the transmission coefficient was close to unity, but it was much less than unity just above the threshold. In this range of energies close to the threshold value the calculations indicated that it was most important to take into consideration the bending motion in the activated state.

Several aspects of the quantum-mechanical problem of reaction rates have been considered by Child. In his first paper[3] he treated the general type of reaction A + BC → AB + C with the atoms constrained to move in a plane and with B having infinite mass. Methods are given for relating the final energy of the product AB to the initial states of the reactants and the impact parameters. The treatment is applied to the reaction K + HI → KI + H on the basis of two hypothetical potential-energy surfaces. One of these has the same form as $\mathcal{U}_{MB}$ employed by Blais and Bunker and by Raff and Karplus (see page 194) for the similar reaction K + CH_3I → KI + CH_3. The other potential-energy surface used by Child has the energy variation smoothed out in such a way that there is no activation barrier. The quantum-mechanical calculations show that the effect of removing the barrier is not only to increase the overall reaction probability but to increase the internal energy of the product KI from 3.3 to 8.3 kcal/mole.

Child[4] has also applied a somewhat similar treatment to the H + H_2 reaction, the motion being confined to two dimensions. Special attention is paid to the correlation between the rotational energies of the reactant

[1] J. Mazur and R. J. Rubin, *J. Chem. Phys.*, **31**:1395 (1959).

[2] E. M. Mortensen and K. S. Pitzer, *Chem. Soc. (London), Spec. Publ.* **16**:57 (1962).

[3] M. S. Child, *Proc. Roy. Soc. (London)*, **A292**:272 (1966).

[4] M. S. Child, *Discussions Faraday Soc.*, **44**:68 (1967).

and product molecules, and it is concluded that the product is formed in the same rotational state as the reactant. The calculations give the range of impact parameters for which reaction occurs, and the rate constant. The expression for the rate constant is identical with that given by activated-complex theory.

Child[1] also explored the effect of a basin at the activation barrier. Quantum-mechanical theory is applied to the general reaction A + BC → AB + C with the atoms constrained to be collinear but with the system free to rotate. A formal potential-energy surface having a basin was employed, and rate constants were calculated over a range of temperature. It is found that the Arrhenius plot is curved, with linearity at the high and low temperature limits. At the high temperature limit the activation energy corresponds to the top of one of the activation barriers, but in the low temperature limit it corresponds to the zero-point level of the stable complex. This behavior is of the same kind as that predicted on the basis of simple tunneling through a basinless potential-energy barrier.

The H + H$_2$ reaction has also been studied quantum mechanically by Karplus and Tang,[2] who used the potential-energy surface developed by Porter and Karplus and employed in the classical calculations of Karplus, Porter, and Sharma (see pages 33 and 160). Karplus and Tang considered two simplified and limiting models; in one, the molecule is considered unperturbed by the incoming atom, and in the other, it interacts adiabatically with the incoming atom. The latter model appeared to be more appropriate. Reaction cross sections were calculated for a range of impact parameters, and the results were found to be very similar to those obtained by Karplus, Porter, and Sharma with a classical treatment. Quantum-mechanical tunneling is therefore not of great importance in this reaction, but the quantum calculations yield a significantly higher energy threshold for reaction than does the classical treatment. This is understandable from the discussion on page 168, where it was pointed out that quantum restrictions in the region of the activated state will limit the amount of vibrational energy available for crossing the barrier. As in the classical treatment, rebound was important over a range of lower energies; this behavior is probably linked to the AB-C repulsion referred to above. As the collision energy increases, however, stripping becomes more important.

[1] M. S. Child, *Mol. Phys.*, **12**:401 (1967).
[2] M. Karplus and K. T. Tang, *Discussions Faraday Soc.*, **44**:56 (1967).

8
Other Theories of Chemical Kinetics

So far in this book we have considered only those theories which treat chemical reactions from the standpoint of motion over a potential-energy surface. During recent years there have been developed a considerable number of alternative kinetic formulations, which do not explicitly regard chemical reaction as a passage over a potential-energy surface. It is difficult to classify these theories in a logical manner, but by and large they fall into one of the following two classes:

1. Theories based on nonequilibrium statistical mechanics. Such theories have been formulated in particular with a view to examining the equilibrium hypothesis inherent in activated-complex theory.
2. Stochastic theories, which treat reaction rates on the basis of the theory of probability, in terms of the theory of the "random walk." These theories are essentially more refined collision theories in which account is taken of the energy levels in the reacting molecules.

At the present time the theories that do not take explicit account of potential-energy surfaces have a somewhat different status from those that do. As we have seen in previous chapters, the theories based on potential-energy surfaces are reaching the stage at which they can give a fairly satisfactory interpretation of the rates of actual reactions, particularly of the simple ones. The other theories, however, are still rather far from being able to do this; to a large extent they have been developed only along formal lines and have not been applied to individual reactions. The main usefulness of these theories has been in providing a fresh insight into the way in which reactions occur and, more specifically, in giving an evaluation, from a different point of view, of the equilibrium hypothesis employed in activated-complex theory.

Because of this more limited applicability of these theories they will not be treated in any detail.

NONEQUILIBRIUM STATISTICAL MECHANICS

Rate theories based on nonequilibrium statistical mechanics have been formulated in particular by Curtiss,[1] Prigogine and coworkers,[2] Present,[3] Mahan,[4] and Yamamoto.[5]

The work of Curtiss and of Prigogine et al. was directed toward examining the equilibrium assumption inherent in activated-complex theory. The treatments were based upon the statistical-mechanical formulations of irreversible processes and made use of particular values for reaction cross sections. The technique was to obtain expressions for the perturbation of the Maxwell-Boltzmann distribution by using various formulations for transition probabilities.

Present's treatment was a reexamination of the same problem with more reasonable values of the reaction cross sections. The reacting molecules were assumed to be spherically symmetrical, and steric factors, internal degrees of freedom, and heats evolved in the reaction were neglected. His conclusion was that if $\varepsilon^*/kT = 5$ (where ε^* is the activation energy per molecule), the simple collision theory, which tacitly assumes equilibrium to exist between reactants and activated complexes, is in error by only 8 percent. For higher values of ε^*/kT the error is less. The error in the equilibrium theories is due to their neglect of the deple-

[1] C. F. Curtiss, *Wisc., Univ., Rept.* CM-476 (1948).

[2] I. Prigogine and E. Xhrouet, *Physika*, **15**:913 (1949); I. Prigogine and M. Mahieu, *Physika*, **16**:51 (1950).

[3] R. D. Present, *J. Chem. Phys.*, **31**:747 (1959).

[4] B. H. Mahan, *J. Chem. Phys.*, **32**:362 (1960).

[5] T. Yamamoto, *J. Chem. Phys.*, **33**:281 (1960).

tion, due to reaction, of the more energetic reactant species; this depletion is only significant if ε^*/kT is small. Present's conclusion is very similar to that obtained with the stochastic theories, as will be seen later in this chapter.

Mahan carried out similar calculations, using the methods of Prigogine and confining his attention to free-radical combination reactions with zero activation energy. He concluded that the equilibrium assumption is justified provided that the radicals are present at a mole fraction of less than 0.1.

Yamamoto based his treatment on a different formulation of the statistical mechanics of irreversible processes, due to Mori[1] and to Kubo et al.[2] De Donder[3] showed that near equilibrium the reaction rate is proportional to the affinity, and Yamamoto expressed the proportionality coefficient as a special type of time-correlation function, in terms of a theory of binary collisions. The rate constant is determined from this proportionality constant, and it is assumed that the same rate constant applies far from equilibrium as close to equilibrium.

STOCHASTIC THEORIES[4]

The term *stochastic* seems first to have been applied to rate theories by Montroll and Shuler,[5] although Kramers[6] and Zwolinski and Eyring[7] used methods of the same kind. A stochastic theory may be defined as one based on the principles of probability. A simple and satisfactory definition was given by Feller:[8] "The terms stochastic processes and random processes are synonyms and cover practically all the theory of probability from coin tossing to harmonic analysis. In practice, the term 'stochastic process' is used mostly when a time parameter is introduced. . . . In stochastic processes the future is never uniquely determined, but we have at least probability relations enabling us to make predictions."

[1] H. Mori, *J. Phys. Soc. Japan*, **11**:1029, 1956.

[2] R. Kubo, M. Yokota, and S. Nakajima, *J. Phys. Soc. Japan*, **12**:1203 (1957).

[3] T. de Donder, *Bull. Classe Sci. Acad. Roy. Belg.*, **7**:197, 205 (1922).

[4] For a general review see D. A. McQuarrie, "Stochastic Approach to Chemical Kinetics," Methuen & Co. Ltd., London, 1967; see also B. Widom, *Science*, **148**:1555 (1965).

[5] E. W. Montroll and K. E. Shuler, *Advan. Chem. Phys.*, **1**:361 (1958).

[6] H. A. Kramers, *Physika*, **7**:284 (1940).

[7] B. Zwolinski and H. Eyring, *J. Am. Chem. Soc.*, **69**:2702 (1947).

[8] A. Feller, "Probability Theory and its Applications," vol. 1, John Wiley & Sons, Inc., New York, 1950.

The model of Kramers   In Kramers' theory the molecules are supposed to become activated through their collisions with other molecules of the surrounding medium, which act as a constant-temperature heat bath. After many exchanges of energy during such collisions a molecule may acquire sufficient energy, the activation energy, to cross over a potential-energy barrier.   This crossing constitutes the chemical reaction, and the rate of crossing is the rate of the reaction.

The interaction of the reactant molecules with the heat bath is analogous to the Brownian motion of the particle in a viscous medium.   The interaction of the reactant with the constant-temperature bath is expressed in terms of a viscosity coefficient η, a large value of η meaning that there is a strong interaction between the reacting molecules and the heat bath.

The rate of reaction is given by the diffusion current over the potential-energy barrier, and the energy distribution of the reacting species along the reaction coordinate is given through the density distribution in momentum space.   Neither the value nor the analytical form of the coefficient of viscosity, η, can be determined from our present knowledge of intermolecular forces.   In the equilibrium theories of kinetics it is not necessary to make use of this quantity, or of anything equivalent, since the properties of the equilibrium state are independent of the method of its establishment.

Because of mathematical difficulties Kramers had to restrict his discussion to the case in which the barrier height is large compared with kT, the mean thermal energy of the molecules, and he was obliged to treat diffusion over the barrier as a quasi-stationary process.   He found that under these conditions the calculated rate is very close to that calculated on the basis of equilibrium theory.   Indeed, for ε^*/kT equal to 10 the agreement is within 10 percent, the agreement coming closer, the larger the ε^*/kT value.   This conclusion is similar to that found by Curtis and Prigogine, as has been seen, and by Montroll and Shuler, as discussed later in this chapter.

Bak[1] has given a theory that is somewhat similar to, but more highly developed than, Kramers' theory, and his conclusions are much the same.

The model of Zwolinski and Eyring   In the treatment by Zwolinski and Eyring[2] the reactants in a chemical reaction are described by one set of quantum states, and the reactant products by another.   These levels are left quite general, not being identified with any particular type of motion. Molecules in the reactant states are postulated to pass, by collision with other molecules, into the product states.

[1] T. Bak, "Contributions to the Theory of Chemical Kinetics," W. A. Benjamin, Inc., New York, 1963.
[2] B. Zwolinski and H. Eyring, *J. Am. Chem. Soc.*, **69**:2702 (1947).

For a first-order reaction the rates of transition between the various reactant and product levels were represented by linear equations of the form

$$\frac{dx_n}{dt} = \sum_m (W_{nm}x_m - x_n W_{mn}) \tag{1}$$

Here x_n is the fraction of molecules in state n at time t, and W_{nm} is the probability per unit of time for a transition, through collision, from state m to state n. Terms with $n = m$ are omitted. There is an equation of the form of Eq. (1) for values of n equal to 0, 1, 2, etc.

The rate coefficients W_{nm} are analogous to the coefficient of viscosity in the Kramers model. In principle they are calculable from the quantum-mechanical theory of collisions, but our present ignorance of intermolecular forces and interactions prevents us from deducing their analytical form or numerical values. In the absence of this information Zwolinski and Eyring assumed certain relations among the various rate constants and assigned plausible numerical values to enough of them that all could be determined.

The general solution of a set of equations of the form of (1) is

$$x_n(t) = \sum_j B_{nj}e^{\lambda_j t} \tag{2}$$

where the λ_j's are the characteristic roots of the matrix of the coefficients B_{nj} obtained by setting x_n equal to $B_n e^{\lambda t}$. Zwolinski and Eyring evaluated (2) numerically for a model involving four energy levels and obtained the time-dependent concentration $x_n(t)$. The rate of reaction, i.e., the rate of passage between energy levels, was then obtained by computing the products $W_{nm}x_m$ and summing these products over the reactant and product levels.

Zwolinski and Eyring also made calculations on the basis of equilibrium theory and found in general that the ratio of nonequilibrium to equilibrium rate is less than unity, the maximum deviation from the equilibrium rate for the four-level model being about 20 percent.

The model of Montroll and Shuler   Zwolinski and Eyring and also Shuler[1] suggested that it would be instructive to extend this four-level model to one of N levels, with a systematic treatment of transitions between the levels. Such an extension was carried out by Montroll and Shuler,[2] who characterized the transitions of the molecules between levels

[1] K. E. Shuler, 5th Symp. Combust., Pittsburgh, **1955**:56–74.

[2] E. W. Montroll and K. E. Shuler, Advan. Chem. Phys., **1**:361 (1958).

as a one-dimensional random walk.   The theory was developed in terms
of the properties of stochastic matrices, whose elements involve the transition probabilities.   The chemical reaction was regarded as the removal
of the reactant molecule from the system when it has reached the
$(N + 1)$st energy level.   The rate of the chemical reaction is then given
by the *mean first-passage time* $\bar{t}$, which is the average time it takes for a
species to pass level N for the first time and so be removed from the
system.

Montroll and Shuler applied their treatment to a unimolecular process.   They considered an ensemble of reactant molecules with quantized
energy levels, immersed in a large excess of gas which acts as a heat bath
at constant temperature T throughout the reaction; this requires that the
heat-bath molecules be present in great excess.   The reactant molecules
are initially in a Maxwell-Boltzmann distribution corresponding to a temperature T_0, such that $T_0 < T$.   As the reactant molecules collide in their
random motion with the heat-bath molecules, they are excited in a step-
wise fashion from one energy level to another until they go past level N,
when they are removed irreversibly from the reaction system.   The transition from level n to level m is characterized by a transition probability
W_{mn}, the magnitude of which depends on the quantum numbers n and
m.   The transitions between levels m and n can be represented as

$$\mathrm{A}_n + \mathrm{M} \underset{W_{nm}}{\overset{W_{mn}}{\rightleftharpoons}} \mathrm{A}_m + \mathrm{M}$$

Since it is assumed that $[\mathrm{A}] \ll [\mathrm{M}]$, collisions between pairs of A molecules are not considered, so that the process of energization is first order
in A; the differential equations governing the rates are therefore linear in
the concentration of the reacting species A.

The stochastic process corresponding to this model of a chemical
reaction is that of a one-dimensional random walk with an absorbing barrier.   In this random walk, the probability per unit of time, W_{mn}, that a
walker will take a step from level n to m is a function of the distance from
the origin (where $n = 0$).   The time-dependent distribution of the reactant molecules among the energy levels $n = 0, 1, 2, \ldots , N$ is then given
by the fraction $x_n(t)$ of walkers who are n levels from the origin at time t.

Montroll and Shuler developed general equations on this basis and
then applied their treatment to the truncated harmonic-oscillator model
shown in Fig. 75.   The truncation allows reaction to occur when the
system passes the Nth level.   The quantum levels are equally spaced, the
energy of the Nth level being taken as $Nh\nu$.   Because of serious difficulties in doing otherwise, they considered only transitions between neighboring quantum levels.   For this model the mean *first-passage time* $\bar{t}$ is

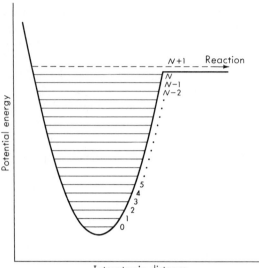

Fig. 75   Energy-level diagram for a harmonic oscillator, with the molecule being removed from the system
at the $(N + 1)$st level.

given approximately (for sufficiently large values of N) by

$$\kappa \bar{t} = \frac{e^{(N+1)h\nu/kT}}{(N+1)(e^{-h\nu/kT} - 1)^2}\left(1 + \frac{e^{-h\nu/kT}}{N(1 - e^{-h\nu/kT})} + \cdots\right) \qquad (3)$$

where κ depends on the coupling between the molecules and the heat bath.
The corresponding expression for equilibrium theory is obtained by setting N equal to infinity and is

$$\kappa \bar{t} = \frac{e^{(N+1)h\nu/kT}}{(N+1)(e^{-h\nu/kT} - 1)^2} \qquad (4)$$

Corrections to equilibrium theory are therefore required when the second
term in (3) is not negligible, i.e., when N is not greater than $e^{-h\nu/kT}$
$(1 - e^{-h\nu/kT})^{-1}$.   The mean first-passage time $\bar{t}$ and the rate of activation,
which is proportional to $\bar{t}^{-1}$, deviate from their equilibrium value by more
than 10 percent when

$$N(1 - e^{-h\nu/kT}) < 10e^{-h\nu/kT} \qquad (5)$$

In the high-temperature limit this corresponds approximately to

$$\frac{Nh\nu}{kT} = \frac{E_{act}}{kT} < 10 \qquad (6)$$

Figure 76 shows the ratio of the equilibrium first-passage time $\bar{t}_{eq}$ to the mean time given by Eq. (3), as a function of N for various values of $h\nu/\mathbf{k}T$. This ratio is the ratio of the reaction rate calculated from the nonequilibrium theory to that calculated from equilibrium theory.

Montroll and Shuler recast Eq. (3) into a form in which it could be applied to experimental results. Their estimate of the frequency factor was found to be

$$A = Z^*P(N + 1)(1 - e^{-h\nu/\mathbf{k}T})^2 \tag{7}$$

where Z^* is the collision number, and P, the transition probability for collision. They found, however, for the rate of activation of iodine molecules by various inert gases, that their calculated rates were too low by several orders of magnitude. The treatment is evidently too simple for it to be able to give reliable rates of reactions.

In spite of this, however, the treatment is an important contribution to reaction-rate theory. Its most valuable feature is that it leads to conclusions about the extent to which the rate of a reaction differs from that calculated on the basis of equilibrium theory, such as by simple collision theory or activated-complex theory. It can be seen from Fig. 76 that the error is about 10 percent if $\varepsilon^*/\mathbf{k}T$ is 10; if $\varepsilon^*/\mathbf{k}T$ is 5, the error is about

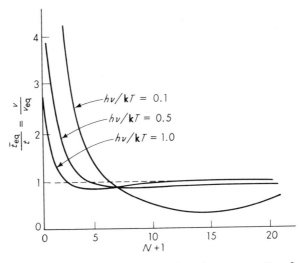

Fig. 76   The ratio of the equilibrium first-passage time $\bar{t}_{eq}$ to the mean time given by the stochastic harmonic-oscillator model [Eq. (3)], as a function of N. This ratio is equal to the ratio of the rate calculated with this model to the rate calculated using the equilibrium assumption.

20 percent, while if ε^*/kT is much greater than 10, the error is negligible. We have seen that similar estimated errors are given by other treatments.

Several serious deficiencies in their treatment were clearly recognized by Montroll and Shuler. One of them was their use of a simple harmonic-oscillator model, in order to simplify the theory. In a real molecule the density of energy levels increases with increasing energy, and there is a corresponding increase in the probability of nonnearest-neighbor transitions. Also, various mathematical approximations are made in the treatment, and it is sometimes difficult to estimate what effect these would have on the final results.

A serious weakness in the treatment is the restriction to transitions between neighboring energy levels. This has the effect of leading to an abnormally slow pathway to activation. Transitions between neighboring levels are the most important, but the totality of the other transitions will make an important contribution to reaction.

In a later paper Shuler and Weiss[1] investigated the dependence of $\bar{t}$ on the initial distribution of molecules, on the height of the barrier, and on the distribution of transition probabilities for non-nearest-neighbor interactions. Their object was to see whether consideration of these factors could reduce the discrepancy between calculated and observed rates. They found that the choice of initial distributions did not significantly affect $\bar{t}$. Allowing transitions between nonnearest neighboring levels, however, significantly increased the calculated rate. Agreement with experiment could, in fact, be obtained if transition probabilities between nonnearest neighbors were taken to be the same as those between nearest neighbors.

Other stochastic treatments   Widom and Bauer[2] developed what is essentially a refined collision theory and applied it to the problem of energy transfer in mixtures of carbon dioxide and water. Instead of the crude hard-sphere collision model, they used a symmetrical Lennard-Jones 6-12 potential for the intermolecular forces, and on a classical basis calculated transition probabilities between vibrational states on collision, taking into account the possibility of conversion into translational energy. Widom[3] applied a similar treatment to a dissociation reaction; the dissociating molecules were considered to be dispersed in an inert gas and to gain their energy by collision with the inert-gas molecules. As in the Montroll-Shuler treatment, the theory was developed in terms of mean passage times. Under simplifying conditions, including the condition that $\varepsilon^* \gg kT$, the rate constant reduced to the form $Ze^{-\varepsilon^*/kT}$.

[1] K. E. Shuler and G. H. Weiss, *J. Chem. Phys.*, **38**:505 (1963).
[2] B. Widom and S. H. Bauer, *J. Chem. Phys.*, **21**:1670 (1953).
[3] B. Widom, *J. Chem. Phys.*, **31**:1387 (1959).

Eliason and Hirschfelder[1] developed what is essentially a generalized theory of bimolecular reactions. Their treatment is based on a formal kinetic theory of polyatomic molecules developed by Wang-Chang and Uhlenbeck,[2] who treated molecules in different internal quantum states, as though they were distinct species, and took into account transitions between these states. This theory was extended by Eliason and Hirschfelder to give a formulation of reaction rates. They discuss the conditions under which their theory reduces to activated-complex theory.

OTHER THEORIES

Other theories of reaction rates will be referred to only very briefly.

A number of workers have developed what are generally known as *statistical* or *phase-space* theories of chemical reactions. These differ from activated-complex theory in several respects. In the first place, the statistical theories are formulated in such a way as to describe the probabilities of possible outcomes of single well-defined collisions, in which (as in the dynamical treatments) the initial conditions are specified exactly; in activated-complex theory, on the other hand, a statistical assembly of reactant molecules must be considered. Secondly, in the statistical theories no assumption need be made as to the lifetime or configuration of the activated state. Thirdly, the statistical theories as formulated at present involve an indirect (sometimes known as *complex*) mechanism, as opposed to the direct mechanism of activated-complex theory. Since most reactions appear to be direct, the statistical theories are of somewhat limited applicability, but they have been valuable in interpreting such reactions as are indirect.

The first formulation of a statistical theory was by Keck,[3] who has reviewed his and other theories of this kind.[4] Other treatments have been given by Light,[5] who has also reviewed his treatments and compared them with experiment,[6] with particular reference to the conditions under which his assumptions may be valid.

[1] M. A. Eliason and J. O. Hirschfelder, *J. Chem. Phys.* **30**:1426 (1959).

[2] C. S. Wang-Chang and G. E. Uhlenbeck, "Transport Phenonena in Polyatomic Molecules," *Mich. Univ. Publication* CM-681 (1951).

[3] J. C. Keck, *J. Chem. Phys.*, **29**:410 (1958); **32**:1035 (1960); *Discussions Faraday Soc.*, **33**:173 (1962).

[4] J. C. Keck, *Advan. Chem. Phys.*, **13**:85 (1967).

[5] J. C. Light, *J. Chem. Phys.*, **40**:3221 (1964); P. Pechukas and J. C. Light, *J. Chem. Phys.*, **42**:3281 (1965); J. C. Light and J. Lin, *J. Chem. Phys.*, **43**:3209 (1965); P. Pechicas, J. C. Light, and C. Rankin, *J. Chem. Phys.*, **44**:794 (1966); J. Lin and J. C. Light, *J. Chem. Phys.*, **45**:2545 (1966).

[6] J. C. Light, *Discussions Faraday Soc.*, **44**:14 (1967).

Name Index

Name Index

Adams, E. P., 42
Airey, J. C., 178
Ajzenberg-Selove, F., 189
Anlauf, K. G., 178, 203
Arrhenius, S., 1, 4, 8
Ashmore, P. G., 85
Aynoneno, P. J., 85

Back, M. H., 8, 26, 108, 150, 151
Bagdasaryan, K. S., 38
Bak, T., 213
Bamford, C. H., 146
Banerjee, M. K., 189
Barker, R. S., 19, 20
Bates, D. R., 59
Bauer, E., 207
Bauer, S. H., 218
Bell, R. P., 60, 62
Bérces, T., 37
Bernstein, R. B., 182, 185, 188, 190, 200
Berry, R. S., 174
Beuhler, R. J., 188
Bigeleisen, J. W., 86, 94
Birely, J. H., 190
Birkenshaw, K., 202
Bishop, D. M., 50, 68, 78
Blais, N. C., 40, 156, 159, 179, 180, 194
Blythe, A. R., 182
Boato, G., 166
Bodenstein, M., 85, 98
Bonner, W. A., 85
Boys, S. F., 20, 21, 22
Brooks, P. R., 188
Brown, J. W., 86
Bruner, B. L., 20, 22–24, 35
Bull, T. H., 184
Bunker, D. L., 40, 117, 145, 156, 157, 159, 179, 180, 194, 205

Butler, S. T., 189
Bywater, S., 84, 86

Caldin, E. F., 62
Carabetta, R. A., 206
Careri, G., 166
Carr, R. W., 148
Cashion, J. K., 31, 178
Chanmugan, J., 85
Charters, P. E., 178
Cherwell, Lord (F. A. Lindemann), 108
Chesick, J. P., 145
Child, M. S., 191, 201, 208, 209
Chiltz, G., 94
Cimoni, A., 166
Clough, P. N., 204
Conroy, H., 20, 22–24, 35
Cook, G. B., 22
Coolidge, A. S., 16, 30
Cooper, W., 149
Coull, J., 149
Coulson, C. A., 59
Crosby, H. J., 85
Cross, P. C., 103
Curtiss, C. F., 211

Datz, S., 184, 185
Decius, J. C., 103
de Donder, T., 212
De Haas, N., 91–93

Eckart, C., 60
Eckling, R., 94
Edmiston, C., 20, 22
Eliason, M., 6, 219
Elliott, C. S., 149

Subject Index

Subject Index

This book was set in Modern by The Maple Press Company and printed on permanent paper and bound by The Maple Press Company. The drawings were done by Engineering-Drafting Company. The editors were James L. Smith and Janet Wagner. Stuart Levine supervised the production.